Owning the TECHNICAL BASELINE for Acquisition Programs in the U.S. Air Force

Committee on Owning the Technical Baseline for Acquisition Programs in the U.S. Air Force: A Study

Air Force Studies Board

Division on Engineering and Physical Sciences

A Report of

The National Academies of
SCIENCES · ENGINEERING · MEDICINE

THE NATIONAL ACADEMIES PRESS
Washington, DC
www.nap.edu

THE NATIONAL ACADEMIES PRESS **500 Fifth Street, NW** **Washington, DC 20001**

This activity was supported by Grant FA9550-14-1-0127 from the U.S. Air Force. Any opinions, findings, conclusions, or recommendations expressed in this publication do not necessarily reflect the view of any organization or agency that provided support for the project.

International Standard Book Number-13: 978-0-309-44903-8
International Standard Book Number-10: 0-309-44903-0
Digital Object Identifier: 10.17226/23631

Copies of this publication are available for sale from the National Academies Press, 500 Fifth Street, NW, Keck 360, Washington, DC 20001; (800) 624-6242 or (202) 334-3313; http://www.nap.edu.

Printed in the United States of America

Suggested citation: National Academies of Sciences, Engineering, and Medicine. 2016. *Owning the Technical Baseline for Acquisition Programs in the U.S. Air Force.* Washington, DC: The National Academies Press. doi: 10.17226/23631.

The National Academies of
SCIENCES · ENGINEERING · MEDICINE

The **National Academy of Sciences** was established in 1863 by an Act of Congress, signed by President Lincoln, as a private, nongovernmental institution to advise the nation on issues related to science and technology. Members are elected by their peers for outstanding contributions to research. Dr. Marcia McNutt is president.

The **National Academy of Engineering** was established in 1964 under the charter of the National Academy of Sciences to bring the practices of engineering to advising the nation. Members are elected by their peers for extraordinary contributions to engineering. Dr. C. D. Mote, Jr., is president.

The **National Academy of Medicine** (formerly the Institute of Medicine) was established in 1970 under the charter of the National Academy of Sciences to advise the nation on medical and health issues. Members are elected by their peers for distinguished contributions to medicine and health. Dr. Victor J. Dzau is president.

The three Academies work together as the **National Academies of Sciences, Engineering, and Medicine** to provide independent, objective analysis and advice to the nation and conduct other activities to solve complex problems and inform public policy decisions. The Academies also encourage education and research, recognize outstanding contributions to knowledge, and increase public understanding in matters of science, engineering, and medicine.

Learn more about the National Academies of Sciences, Engineering, and Medicine at **www.national-academies.org.**

The National Academies of
SCIENCES · ENGINEERING · MEDICINE

Reports document the evidence-based consensus of an authoring committee of experts. Reports typically include findings, conclusions, and recommendations based on information gathered by the committee and committee deliberations. Reports are peer reviewed and are approved by the National Academies of Sciences, Engineering, and Medicine.

Proceedings chronicle the presentations and discussions at a workshop, symposium, or other convening event. The statements and opinions contained in proceedings are those of the participants and have not been endorsed by other participants, the planning committee, or the National Academies of Sciences, Engineering, and Medicine.

For information about other products and activities of the Academies, please visit nationalacademies.org/whatwedo.

COMMITTEE ON OWNING THE TECHNICAL BASELINE FOR ACQUISITION PROGRAMS IN THE U.S. AIR FORCE: A STUDY

HENRY A. OBERING III, Booz Allen Hamilton, *Chair*
LAWRENCE J. DELANEY, Independent Consultant, *Vice Chair*
DONALD R. ERBSCHLOE, Technical Consultant
MILLARD S. FIREBAUGH, University of Maryland
MICHAEL D. GRIFFIN, Schafer Corporation
GARY A. KYLE, Persistent Agility, Inc.
THOMAS L. MAXWELL, Independent Aerospace Consultant
SUE C. PAYTON, SCI Aerospace, Inc.
RICHARD T. ROCA, Applied Physics Laboratory, Johns Hopkins University
WILLIAM J. STRICKLAND, HumRRO
DEBORAH L. WESTPHAL, Toffler Associates
REBECCA WINSTON, Winston Strategic Management Consulting

Staff

JOAN FULLER, Director, Air Force Studies Board
ANDREW J. KREEGER, Study Director
DIONNA C. ALI, Research Assistant
STEVEN DARBES, Research Assistant
ADRIANNA HARGROVE, Senior Program Assistant/Financial Assistant
CHRIS JONES, Financial Manager
MARGUERITE E. SCHNEIDER, Administrative Coordinator
JOAN ZIMMERMANN, Consultant

Foreword

After my early operational flying assignments, I began my U.S. Air Force (USAF) acquisition career in 1980 coming out of graduate school. At that time, USAF acquisition was considered the gold standard in the Department of Defense (DoD). The culture was one of innovation and risk taking, producing systems that performed in Operation Desert Storm, including the stealth fighter/bomber, precision-guided weapons, and so on. I personally experienced the quality of the USAF acquisition community when I served as the executive officer to the Assistant Secretary of the Air Force (Acquisition) in the early 1990s. The vast majority of the USAF program managers and the System Program Office teams that briefed their programs through the DoD oversight chain clearly "owned" the technical baseline of their systems, were knowledgeable about the industry, and were authoritative and accountable for their performance.

When I compare what I observed then to what the committee heard in its interviews for this study, it is clear that the USAF acquisition culture of innovation and risk taking has eroded. There are definitely very professional program executive officers (PEOs) and program managers (PMs) who are performing well in spite of the barriers to their success, which the committee discusses in this report. But it is also clear that the USAF needs to take critical steps to emphasize the value of the USAF acquisition professional; reinforce the PM's authority and accountability; clarify the role of the contracting officer with the PM; strengthen the technical expertise of the acquisition workforce; and continue to knock down barriers as they arise. This is especially important in light of the ever shorter timeframes within which the Air Force needs to develop and deploy warfighting capabilities to meet rapidly emerging threats.

This study was not meant to be another attempt at acquisition reform. The committee wanted to stay focused on concrete and achievable steps that the USAF could take to strengthen its ability to produce and maintain weapons systems in a more timely and cost-effective manner. We are truly fighting tomorrow's wars today in our system program offices, and we should make sure that our acquisition warfighters have every available weapon in their arsenal.

Lt. Gen. (USAF, Ret.) Henry A. "Trey" Obering III, *Chair*
Committee on Owning the Technical Baseline for
Acquisition Programs in the U.S. Air Force: A Study

Preface

This study was conducted as a follow-on activity to the National Research Council[1] workshop "Owning the Technical Baseline for Acquisition Programs in the U.S. Air Force." The workshop created the framework and foundational information this study utilized to explore the important topic of the U.S. Air Force (USAF) owning the technical baseline in its acquisition programs. During the workshop the topics of programs, leadership and culture, workforce, contracting, and funding were identified as fundamental components of owning the technical baseline. The workshop and its subsequent report,[2] published in 2015, were leveraged as a reference document and as a foundation for this study's approach to address issues identified during the workshop.

STATEMENT OF TASK AND STUDY APPROACH

The Air Force Studies Board (AFSB) of the National Academies of Sciences, Engineering, and Medicine was asked by the Assistant Secretary of the Air Force for Science, Technology and Engineering to build on the work of the aforementioned workshop and provide recommendations to improve the USAF's capabilities to

[1] Effective July 1, 2015, the institution is called the National Academies of Sciences, Engineering, and Medicine. References in this report to the National Research Council are used in a historical context for identifying programs prior to that date.

[2] National Research Council, *Owning the Technical Baseline for Acquisition Programs in the U.S. Air Force: A Workshop Report*, The National Academies Press, Washington, D.C., 2015.

own the technical baseline for its acquisition programs. The National Academies approved the statement of task for this study in August 2015[3] and in October 2015 appointed the Committee on Owning the Technical Baseline for Acquisition Programs in the U.S. Air Force: A Study.[4] The committee was asked to address the following questions:

1. What is the strategic value to the U.S. Air Force in properly controlling, as well as the risk of not controlling, the technical baselines of its programs?
2. How do others (e.g., services, government agencies, and commercial industry) control technical baselines and what are the most promising mechanisms for potential application within the U.S. Air Force?
3. Are there ways to remove or remediate barriers across the U.S. Air Force, such as barriers identified in *Owning the Technical Baseline for Acquisition Programs in the U.S. Air Force: A Workshop Report* to properly control the technical baselines of future programs?
4. How can the U.S. Air Force assess and adopt any identified methods for controlling the technical baselines across its acquisition programs?

To address these questions, the committee held three data-gathering meetings, which included face-to-face and telephone interviews, from January through March 2016 to review the information presented and discussed during the workshop, independently research the topic, conduct interviews with experts, identify key findings, and develop recommendations. A fourth meeting was held in May 2016 for the committee to write the report. Throughout the meetings, the committee met with current and former senior personnel from the USAF, the Navy, Department of Defense (DoD), National Aeronautics and Space Administration (NASA), National Nuclear Security Administration (NNSA), Missile Defense Agency (MDA), Defense Acquisition University (DAU), and industry.[5] In this context, the committee aimed to identify the strategic and operational value to the USAF of properly controlling, or "owning," the technical baselines of its programs by investigating how other services, government agencies, and industry control technical baselines and qualitatively measure the success of their control. The committee sought to identify and recommend methods to remove or remediate barriers to owning the technical baseline that exist across the USAF. These recommendations were developed by examining how the programs, leadership and culture, workforce, contracting, and funding all play roles in the decisions regarding programs and controlling the technical baseline.

[3] Appendix A provides the statement of task for this study.

[4] Appendix B provides short biographies of the committee members.

[5] Appendix C provides a list of meetings and speakers.

Acknowledgment of Reviewers

This report has been reviewed in draft form by individuals chosen for their diverse perspectives and technical expertise. The purpose of this independent review is to provide candid and critical comments that will assist the institution in making its published report as sound as possible and to ensure that the report meets institutional standards for objectivity, evidence, and responsiveness to the study charge. The review comments and draft manuscript remain confidential to protect the integrity of the deliberative process. We wish to thank the following individuals for their review of this report:

R. Stephen Berry, University of Chicago,
Holly Dunlap, Raytheon Company,
Gordon R. England, PFP Cybersecurity,
John M. Griffin, Griffin Consulting,
Ronald T. Kadish, Booz Allen Hamilton,
William A. LaPlante, The MITRE Corporation,
Paul D. Nielsen, Carnegie Mellon University,
Jon S. Ogg, AeroVise, LLC, and
Starnes E. Walker, University of Delaware.

Although the reviewers listed above have provided many constructive comments and suggestions, they were not asked to endorse the conclusions or recommendations, nor did they see the final draft of the report before its release. The review of this report was overseen by Chris G. Whipple, ENVIRON (retired), who was responsible for making certain that independent examination of this report was carried out in accordance with institutional procedures and that all review comments were carefully considered. Responsibility for the final content of this report rests entirely with the authoring committee and the institution.

Contents

APPENDIXES

Acronyms

A&AS	Advisory and Assistance Services
AFFARS	Air Force Federal Acquisition Regulation Supplement
AFLCMC	Air Force Life Cycle Management Center
AFNWC	Air Force Nuclear Weapons Center
AFOCD	Air Force Officer Classification Directory
AFSB	Air Force Studies Board
AFWCF	Air Force Working Capital Fund
AWQI	Acquisition Workforce Qualification Initiative
BBP	better buying power
CLoA	contracting line of authority
CO	contracting officer
DAE	Defense Acquisition Executive
DAU	Defense Acquisition University
DAWIA	Defense Acquisition Workforce Improvement Act
DCGS	Distributed Common Ground System
DoD	Department of Defense
DoDI	Department of Defense Instruction
EPASS	Engineering, Professional, and Administrative Support Services

EWI	Education with Industry
FAR	Federal Acquisition Regulation
GAO	Government Accountability Office
GPS	Global Positioning System
GSA	General Services Administration
KPP	key performance parameter
LPTA	lowest price, technically acceptable
MAC	multiple award contract
MDA	Missile Defense Agency
MDAP	major defense acquisition program
NAICS	North American Industry Classification System
NASA	National Aeronautics and Space Administration
NNSA	National Nuclear Security Administration
NRC	National Research Council
NRO	National Reconnaissance Office
O&M	operations and maintenance
OASIS	One Acquisition Solution for Integrated Services
OCX	operational control segment
OEM	original equipment manufacturer
OSD	Office of the Secretary of Defense
OTB	owning the technical baseline
PEO	program executive officer
PM	program manager
PM-LOA	program management line of authority
PMO	Program Management Office
PPBE	planning, programming, budget, and execution
RDT&E	research, development, testing, and evaluation
RFP	request for proposal
ROTC	Reserve Officers' Training Corps
S&E	science and engineering
SAF/AQ	Assistant Secretary of the Air Force for Acquisition

SAF/FM	Assistant Secretary of the Air Force for Financial Management and Comptroller
SAF/GC	Assistant Secretary of the Air Force General Council
SAF/IE	Assistant Secretary of the Air Force for Installations, Environment, and Logistics
SAF/MR	Assistant Secretary of the Air Force for Manpower and Reserve Affairs
SBIRS	space-based infrared system
SE&I	systems engineering and integration
SecAF	Secretary of the Air Force
SMC	Space and Missile Systems Center
SPO	Systems Program Office
STEM	science, technology, engineering, and mathematics
TA/LEP	technically acceptable, lowest evaluated price
TRL	technology readiness level
TSPR	total system performance responsibility
TTP	tactics, techniques, and procedure
USAF	U.S. Air Force
USD (AT&L)	Under Secretary of Defense for Acquisition, Technology, and Logistics
WSARA	Weapons Systems Acquisition Reform Act

Summary

We are fighting tomorrow's wars in today's program development offices.
—Lt. Gen. (USAF, Ret.) Henry A. "Trey" Obering III

INTRODUCTION

Entering the 1990s the U.S. Air Force (USAF) was widely recognized as being a premier technical acquisition enterprise.[1] In the decades leading up to the 1990s, the USAF possessed a cadre of technical experts who were well respected by industry owing to their extensive experience with weapon system development; their knowledge of their disciplines; their understanding of the greater system, or system of systems, in which the product was to operate; and their possession of the charter and authority to make trades among the technical requirements necessary to optimize the product within the cost and schedule constraints.

Beginning in the mid-1990s, during the post-Cold War drawdown of the services, the USAF reacted to shifts in the strategic and budgetary environment. These reactions had the cumulative effect of distancing the Air Force from its close technical oversight of the contractors who were building the Air Force's highly complex systems. The indiscriminate application of the total system performance responsibility (TSPR) acquisition methodology and varied interpretations of its intent; reductions in the civil service technical workforce; over-emphasis on contract price

[1] G.E. Christle, D. Davis, G. Porter, *CNA Independent Assessment: Air Force Acquisition*, Center for Naval Analyses, Alexandria Va., 2009, p. 1.

as an award criterion; and reactions to ethics violations with respect to government employee–contractor relationships were among the factors that caused the USAF to lose its position as a highly capable and professional enterprise in defense acquisition. TSPR was an overt move by senior Department of Defense (DoD) and Air Force personnel to have Air Force acquisition professionals remove themselves from the daily involvement with their industry counterparts. There was a belief by some that USAF engineers should only be engaged in developing the specifications and should then step back from program execution. This view both supported and was necessitated by reductions in Air Force uniformed and civilian staff with extensive technical backgrounds. These trends were exacerbated in some cases by the use of contract price as the dominant discriminator in determining contract awards, which has led to the all-too-frequent use of lowest price, technically acceptable (LPTA) awards.[2] Taken together, the cumulative effect of these trends has been the USAF's broad and deep loss of its technical baseline, which later contributed to many of the cost and schedule overruns in Air Force acquisition programs.

OWNING THE TECHNICAL BASELINE

New weapons systems require enormous investments in systems and people, as well as tactics, techniques, and procedures (TTPs). The system itself needs to be well understood throughout its life cycle, from concept formulation through operational use. Modern warfighting systems operate within a system of systems, and the complex web of interfaces and integration nodes among the systems also requires planning, investment, and support. The trade-offs and decisions required for optimal planning and execution of these advanced systems can only be made by the entity—in this case the USAF—responsible for the integration of all elements across the entire system and their eventual use to defend U.S. interests.

The term "owning the technical baseline" (OTB) has gained traction among USAF acquisition professionals and leaders; however, the term has not achieved universal recognition, understanding, or implementation. Within the broader DoD acquisition community, the term "ownership" is sometimes mistakenly construed as referring to the legal ownership of technical data rights. This fails to acknowledge that meaningful ownership, in terms of owning the technical baseline, is a much broader concept. Linking the term "ownership" with technical data rights has led to concerns on the part of industry over the protection of intellectual property rights.

[2] The U.S. Air Force's KC-X Program is one example where the Air Force relied on LPTA evaluation criteria. Discussion on the KC-X Program can be found in *The DoD's Use of Lowest Price Technically Acceptable (LPTA) Price Selection.* The full reference is J.S. Gansler, L.H. Harrington, and W. Lucyshyn, *The DoD's Use of Lowest Price Technically Acceptable (LPTA) Price Selection,* UMD-CM-13-098, revised September 2013, p. 10.

While such protection is indeed an important issue, the technical baseline can be owned without encroaching on technical data rights within the current guidelines surrounding intellectual property. Most important, the word "ownership" refers to an acceptance by the service, and by the relevant individuals in the service, of certain fundamental responsibilities for outcomes. These responsibilities cannot be farmed out. As Admiral Hyman G. Rickover put it,[3]

> Responsibility is a unique concept. . . . You may share it with others, but your portion is not diminished. You may delegate it, but it is still with you. . . . If responsibility is rightfully yours, no evasion, or ignorance or passing the blame can shift the burden to someone else. Unless you can point your finger at the man who is responsible when something goes wrong, then you have never had anyone really responsible.

In order to discuss owning the technical baseline, it is important to begin with definitions. The report of the workshop, *Owning the Technical Baseline for Acquisition Programs in the U.S. Air Force: A Workshop Report* established the following definitions for owning the technical baseline:[4]

> *Technical baseline*: Data and information that provide the program office knowledge to establish, trade off, verify, change, accept, and sustain functional capabilities, design characteristics, affordability, schedule, and quantified performance parameters at the chosen level of the system hierarchy.
>
> *Owning the technical baseline*: Air Force program managers and personnel have sufficient technical knowledge of their engineering development programs to enable program success by *making informed, timely, and independent decisions*[5] to manage cost, schedule, and performance risk while ensuring disciplined program execution. Owning the technical baseline allows the Air Force to respond knowledgeably and have minimal disruption to mission success. [emphasis added]

Owning the technical baseline allows the government acquisition team to manage and respond knowledgeably and effectively to systems development, operations, and execution, thereby avoiding technical and other programmatic barriers to mission success. Additionally, owning the technical baseline ensures that government personnel understand the user requirements, why a particular design and its various features have been selected over competing designs, and what the options are to pursue alternative paths to the final product given unanticipated cost,

---

[3] D. Oliver, *Against the Tide: Rickover's Leadership Principles and the Rise of the Nuclear Navy*, Naval Institute Press, 2014, p. 31.

[4] National Research Council, *Owning the Technical Baseline for Acquisition Programs in the U.S. Air Force: A Workshop Report*, The National Academies Press, Washington, D.C., 2015.

[5] Decisions made by the government team should be consistent with the terms and conditions contained in associated contracts.

schedule, and performance challenges. Per the definition of the technical baseline, ownership occurs at the chosen level of system hierarchy. There is a range of technical complexity in programs within the Air Force that drives the need for employing varying levels of technical expertise and knowledge. Acquisition of an off-the-shelf commercial item does not require the same technical expertise to make informed, timely, and independent decisions as do programs that require the design, development, and production of complex weapons systems.

In May 2009, the Secretary of the Air Force and the Air Force chief of staff issued a memorandum stating that "the United States Air Force is committed to recapturing acquisition excellence by rebuilding an Air Force acquisition culture that delivers products and services as promised—on time, within budget, and in compliance with all laws, policies, and regulations.[6]" It is in this context, regaining acquisition excellence, that owning the technical baseline needs to be treated as more than a process or checklist; it needs to be viewed as the result of a holistic, consistent, connected set of technical, business, human capital, and mission strategies and practices.[7]

This report comprises three chapters. Chapter 1 provides context for the study. It includes an overview of the OTB workshop, key observations from that workshop, and the current environment within the Air Force acquisition community related to owning the technical baseline of USAF acquisition programs. Chapter 2 addresses the first item in the statement of task and discusses the strategic value to the Air Force of owning the technical baseline and the risk of not owning it. Additionally, Chapter 2 addresses the second item in the statement of task by highlighting key aspects of how agencies other than the Air Force own the technical baseline for their acquisition programs. Lastly, Chapter 3 addresses the third and fourth items in the statement of task by identifying specific barriers to owning the technical baseline for the Air Force and making recommendations to help guide the Air Force in overcoming those barriers. Each recommendation represents an identified method for the Air Force to adopt to assist in owning the technical baseline of its acquisition programs. The discussion prior to each recommendation in Chapter 3 is provided to assist the Air Force in assessing the identified methods.

---

[6] U.S. Air Force, *Acquisition Improvement Plan*, May 4, 2009, http://www.dodbuzz.com/wp-content/uploads/2009/05/acquisition-improvement-plan-4-may-09.pdf.

[7] Hon. Stan Soloway, president and CEO, Professional Services Council, interview with the committee on January 14, 2016.

RECOMMENDATIONS

The loss of the technical baseline in the USAF began with decisions made by leadership, and it will take the commitment of leadership to enable the Air Force to reestablish its once hard-earned reputation of technical proficiency and acquisition excellence. The Committee on Owning the Technical Baseline for Acquisition Programs in the U.S. Air Force: A Study has formulated a series of recommendations that should assist the Air Force in taking appropriate ownership of the technical baseline and thereby continuing to regain its reputation for excellence in acquisition.

RECOMMENDATION 1: The Secretary of the Air Force should investigate why the position of Assistant Secretary of the Air Force for Acquisition is in an acting or vacant status more frequently than other Air Force Assistant Secretary positions. This investigation should consider how the Air Force, along with other Services and government agencies, fills similar critical positions and should focus on identifying best practices for implementation. The Assistant Secretary of the Air Force for Acquisition position should not be vacant for any extended period of time, and the use of an acting individual should be minimized. Furthermore, in order to attract competitive talent, the Air Force should ensure that it does not impose any additional restrictions beyond those required by law, especially relative to the post-employment period, for the position of Assistant Secretary of the Air Force for Acquisition.

RECOMMENDATION 2: Air Force senior leaders should define, develop, and execute a strategy that balances risk and reward from a program implementation viewpoint, fosters a learning environment characterized by healthy tension and debate, and actively rewards acquisition personnel that regularly find a "pathway to yes." A risk-tolerant acquisition culture, in concert with a sense of urgency, is critical to agile and timely acquisition for the Air Force to maintain its advantage against rapidly evolving threats. Significant attention should be given to the proliferation and acceptance of this crucial change. The strategy should include the following at a minimum:

- Establishing an education and training program to promote and develop a risk-tolerant culture that includes the use of current and former experienced acquisition professionals to provide guidance and mentorship.
- Encouraging the pursuit of more reasonable interpretations of policy and process flexibility to more efficiently accomplish program goals while maintaining compliance.
- Assuring that logical and reasonable deviations from policy or requirements can be expeditiously pursued by empowered acquisition personnel.

In addition to changes that provide consistent tenancy in the Assistant Secretary of the Air Force for Acquisition and a strategy to proliferate a more risk-tolerant acquisition culture, changes in how the Air Force manages its workforce are necessary.

RECOMMENDATION 3: The Air Force should continue and complete its efforts to determine which current programs should own the technical baseline and develop staffing standards to determine the proper mix and number of military and civilian engineers required to own the technical baseline for those programs. Criteria should be established for when the Air Force should own the technical baseline as opposed to having knowledge of the baseline as technical integrator or interface systems reviewer. The decision to own the technical baseline for future programs should be included in the acquisition milestone protocol as gated decision points. Additionally, the Air Force should develop methods to measure whether or not selected programs have successfully achieved, and are maintaining, ownership of the technical baseline. Cost overruns, schedule delays, and unidentified, or incorrectly identified, key performance parameters (KPPs) are potential measurement points.

RECOMMENDATION 4: The Air Force should review, and make appropriate changes to, current assignment policies and practices for the acquisition workforce to reduce turnover and attrition and increase succession and transition planning; should invest in a more structured mentoring program across the acquisition workforce to increase the sharing of best practices; and should ensure that the career management system for the acquisition workforce be charged with providing appropriate educational opportunities, training, and industrial experiences to acquisition personnel. The intent of the review should be to create strong career paths for acquisition personnel reflecting the critical value of acquisition to future Air Force operations.

RECOMMENDATION 5: Air Force leadership should, in concert with its current activities, ensure that there is necessary guidance and governance for the currency of appropriate skills of the acquisition workforce at all levels. This must include, but is not limited to, emphasis on the criticality to program success of technically educated and technically experienced program managers. Additionally, the Air Force should prioritize education and experience in industry, recognize its importance to the development of competent acquisition personnel, and increase the opportunities for members of the acquisition workforce to gain this education and experience.

RECOMMENDATION 6: The Air Force should establish, select, and equip a dedicated line of program acquisition officers, selected from a defined science, technology, engineering, and mathematics (STEM)-intensive career path in the uniformed services. This dedicated line of program acquisition officers would be similar in intent, education, and experience to the Navy's engineering and aeronautical engineering duty officers. Additionally, a robust career path for USAF civilian engineers and program managers should be established that supports their critical importance to the successful execution of acquisition programs through ownership of the technical baseline. Program managers should generally be selected from the engineering and technical workforce.

RECOMMENDATION 7: Air Force Life Cycle Management Center (AFLCMC) leaders should work with the Engineering, Professional, and Administrative Support Services (EPASS) program management office to put in place a rigorous requirements definition process such that specific technical requirements and criteria are approved by the program manager and that contractor personnel align with those requirements to meet the needs of the program. Application of AFLCMC's technically acceptable, lowest evaluated price (TA/LEP) approach should be a secondary consideration to meeting the requirement and delivering customer value.

As the committee found during the Owning the Technical Baseline Workshop and learned from the interviews conducted during the study, the authorities and accountabilities of the program manager (PM) as they relate to the authorities and accountabilities of the contracting officer (CO) are currently causing tension and often negatively impacting the effectiveness of the USAF acquisition team. Contracting support is key to owning the technical baseline, and changes are needed for future success in Air Force acquisition programs.

RECOMMENDATION 8: The Air Force should issue a guidance memorandum that clearly specifies the lines of authority and accountability for all members of the government acquisition team. This memorandum should clarify and reinforce PM authorities and responsibilities as well as specify CO responsibilities, as part of the government acquisition team, in relation to the PM. Specifically, all functional entities should provide the PM with the support necessary to attain program success. All members of the government acquisition team should be measured based on program success while complying with the law. Additionally, the Air Force should revise the Air Force Federal Acquisition Regulation Supplement (AFFARS) to make it clear that Air Force program executive officers (PEOs) and PMs, or their designated representatives, are mandatory participants in business clearance and contract clearance sessions. Program management and contracting personnel should be trained in implementation of the guidance.

RECOMMENDATION 9: Contracting professionals' appraisals should have appropriate objectives and metrics tied directly to the program office or organization's mission success. The PEO or the PM or their designee should be required to provide written performance input to the contracting professionals' annual appraisals. Contracting professionals should engage with the program office and be well trained and experienced with their accountability and responsibility for delivering support to the assigned Air Force organization and mission.

RECOMMENDATION 10: The Assistant Secretary of the Air Force for Acquisition should clarify the criteria for use of the lowest price, technically acceptable (LPTA) methodology and ensure there are avenues for the government acquisition team to discuss its appropriateness for meeting mission requirements. LPTA should not be applied to complex, multiyear, multidiscipline programs or knowledge-based service contracts that require high-end acquisition and technical talent. A decision to use LPTA should depend on clear and unambiguous requirements, underlying market research, and relevant information acquired during government and contractor interactions, such as "industry days." If there is a requirement that demands special treatment, the case should be made in the requirements definition, acquisition strategy, and pre-request for proposal (RFP) activity.

In a constrained budget atmosphere, the efficient use of available funds to support weapons systems is paramount to meeting mission requirements. USAF leaders have recognized this need and have already begun to employ more flexible means of funding USAF's acquisition staff.

RECOMMENDATION 11: The Air Force should complete the shift from operations and maintenance (O&M) funds to research, development, testing and evaluation (RDT&E) funds for funding acquisition staff. Additionally, the Assistant Secretary of the Air Force for Acquisition should require PMs to include in their program financial plan such a budget, as necessary, to fully fund the in-house technical effort.

It is crucial that the Air Force fully implement these recommendations. Enacting only a subset of the recommendations contained in this report will hinder the Air Force's ability to own the technical baseline and regain its acquisition excellence.

BOTTOM LINE

Tomorrow's wars are being fought in today's program development offices. To win those wars, urgent and dramatic steps are needed to ensure that the Air Force removes several major barriers to success. Owning the technical baseline is a critical component of the Air Force's ability to regain and maintain acquisition excellence. There are definitely very professional program executive officers (PEOs) and program managers (PMs) who are performing well in spite of the barriers, but it is clear that the USAF needs to take immediate steps to emphasize the value of its acquisition professionals, ensure sustained leadership within the acquisition community, reinforce the PM's authority and accountability, clarify the role of the contracting officer vis-à-vis the PM, strengthen and expand the technical knowledge base and expertise of the acquisition workforce, and continue to eliminate barriers and avoid creating new ones. These necessary steps for owning the technical baseline are especially important in light of the shorter and shorter time frames within which the Air Force needs to develop and deploy warfighting capabilities to meet rapidly emerging and changing threats.

1

Context of the Study

While there are examples of successful weapon systems acquisition programs within the U.S. Air Force (USAF), many of the programs are still incurring cost growth, schedule delays, and performance problems. The USAF now faces serious challenges in acquiring and maintaining its weapons systems as it strives to maintain its current programs; add new capabilities to counter evolving threats; and reduce its overall program expenditures. The federal budget cycle, characterized by short and volatile planning horizons, also places serious hurdles in the path of effective planning and execution of acquisition programs and inventory maintenance.

In 2009, in response to several events, including contract protests and budget overruns, and the subsequent reports issued by the Government Accountability Office (GAO), the Secretary of the Air Force (SecAF) and the Assistant Secretary of the Air Force for Acquisition (SAF/AQ) met with GAO leadership to discuss the strengths and weaknesses of the Air Force acquisition process.[1] At that time, the Air Force leaders also requested that the Center for Naval Analyses conduct an independent assessment of Air Force acquisition. These assessments concluded with a summary of concerns in five critical areas:[2]

[1] Office of the Assistant Secretary of the Air Force (Acquisition), "Acquisition Improvement Plan," May 4, 2009, p. 1, http://www.dodbuzz.com/wp-content/uploads/2009/05/acquisition-improvement-plan-4-may-09.pdf.

[2] Ibid, p. 2.

1. Degraded training, experience, and quantity of the acquisition workforce;
2. Overstated and unstable requirements that are difficult to evaluate during source selection;
3. Under-budgeted programs, changing of budgets without acknowledging impacts on program execution, and inadequate contractor cost discipline;
4. Incomplete source selection training that has lacked "lessons learned" from the current acquisition environment, and delegation of decisions on leadership and team assignments for major defense acquisition programs (MDAPs) source selections too low; and
5. Unclear and cumbersome internal Air Force organization for acquisition and program executive officer (PEO) oversight.

Numerous studies, papers, and public commentary have repeatedly raised similar concerns over the last two decades. Additionally, the committee found throughout the study process that many of these concerns appear to continue to plague Air Force acquisition programs today. The need to reduce cost overruns and schedule delays and to remedy performance issues has helped push the discussion in the Air Force acquisition community toward the need to "own the technical baseline," a concept that, in combination with corrections to the five critical areas above, is well aligned with the USAF desire to reacquire its lost acquisition excellence.

OVERVIEW OF THE WORKSHOP ON OWNING THE TECHNICAL BASELINE

In October 2014, under the auspices of the Air Force Studies Board, the National Research Council (NRC)[3] appointed the Committee on Owning the Technical Baseline in the U.S. Air Force: A Workshop. The committee planned and participated in the workshop and prepared the report *Owning the Technical Baseline for Acquisition Programs in the U.S. Air Force: A Workshop Report.* The workshop was conducted at the request of the SAF/AQ, and the workshop committee was asked to address the following statement of task:[4]

> 1. Identify the essential elements of the technical baseline that would benefit from realignment under Air Force or government ownership, and the value to the Air Force of regaining ownership under its design capture process of the future.

[3] Effective July 1, 2015, the institution is called the National Academies of Sciences, Engineering, and Medicine. References in this report to the National Research Council (NRC) are used in a historical context to refer to activities before July 1.

[4] National Research Council (NRC), *Owning the Technical Baseline for Acquisition Programs in the U.S. Air Force: A Workshop Report*, The National Academies Press, Washington, D.C., 2015, p. 2.

2. Identify the barriers that must be addressed for the Air Force to regain technical baseline control to include workforce, policy and process, funding, culture, contracts, and other factors.
3. Provide terms of reference for a possible follow-on study to explore the issues and make recommendations required to implement and institutionalize the technical baseline concept, and possibly prototype the concept on a demonstration program for lessons learned.

The context for the workshop, as best described by the report itself, was as follows:

> A number of observers have argued that for years, beginning in the mid-1990s, the U.S. Air Force ceded control, active oversight, and in-depth understanding of the technical baselines for weapon systems to defense prime contractors with negative consequences that included (1) loss of ability to perform independent technical analysis, (2) loss of ability to validate defense contractor technical decisions and conclusions, (3) atrophy of the engineering workforce competency, (4) decrease in the ability to attract and retain top engineering talent due to hands-off engineering, (5) decreased ability to control costs, and (6) a reluctance by industry to share detailed, proprietary technical data for fear of transfer to competitors.[5]

The workshop report summarized individual committee members' observations, based on presentations received from 35 Department of Defense (DoD) program managers (PMs) and current and senior leaders. The purpose and charge of such workshop committees is not, however, to reach consensus on conclusions, findings, or recommendations.

KEY OBSERVATIONS FROM THE WORKSHOP

The Owning the Technical Baseline Workshop consisted of three 2-day sessions. Throughout those sessions the workshop committee heard from numerous experts, and several recurring topics emerged from those discussions. Those recurring topics, contained in Box 1.1, created a basis for this committee's approach to further investigating a construct for the USAF to own its technical baseline.[6]

CURRENT ENVIRONMENT FOR ACQUISITION PROGRAMS IN THE U.S. AIR FORCE

The current environment for USAF acquisition programs is strongly influenced by the overall USAF budget environment, which is complex and challenging. The combination of a decrease in funding and the drawdown of forces at home and abroad has placed greater demand on Air Force assets. This greater demand has created obstacles and dilemmas with regard to trade-offs between meeting the

[5] Ibid., p. 1.
[6] Ibid., pp. 3 and 4.

BOX 1.1
Recurring Topics from the Owning the Technical Baseline Workshop

Programs: According to several workshop participants, the most important issue with respect to managing the technical baseline of a weapon system was for the PEO and PM to be able to oversee and manage the baseline with accountability, authority, and responsibility.

Leadership and Culture: In the opinion of at least two participants, it was essential that senior Air Force leadership make clear to all functional leaders supporting acquisition that the Air Force highly values technically trained and competent acquisition and engineering personnel.

Workforce: In the view of three participants, continuity, longevity, and mentoring in the engineering and technical fields, including a succession pipeline, were crucial for success of a program. They argued that the Air Force needed to implement a formal, robust, and credible training and mentoring program to (1) transfer knowledge to upcoming acquisition professionals and (2) develop demonstrable business acumen in the acquisition workforce.

Funding: According to several participants, the limitations of operations and maintenance (O&M) funding and the inability to use research, development, testing, and evaluation (RDT&E) funding for hiring, retaining, and training the technical acquisition workforce created barriers to success. In their view, this lack of adequate and timely funding limited the ability of acquisition center functional leads from shaping the workforce to meet the demands for knowledgeable and experienced technical talent.

Contracting: In the opinion of two participants, it was essential that contracts reflect the proper level of government technical and business engagement to include oversight, insight, data rights, and intellectual property consistent with the program's life cycle acquisition strategy. This should start at the earliest phases of the program and include a maintenance or sustainment approach for owning the technical baseline over the life cycle of the program.

needs of current warfighting commitments and satisfying long-term U.S. national security requirements.[7]

Aging aircraft, the increasing costs of operation and maintenance, and the growing cost of personnel—all of which are occurring in an era of rapid force buildup of fifth-generation systems by potential adversaries—amplify and underscore the need for capable and efficient acquisition of technically superior systems by the Air Force acquisition team. Based on the many studies and subsequent

[7] U.S. Air Force, Deputy Assistant Secretary for Budget (SAF/FMB), *United States Air Force Fiscal Year 2016 Budget Overview*, 2015, Foreword.

reports,[8] the Air Force has invested much time and funding to understand the current condition of its acquisition workforce. It is generally recognized that the end of the Cold War caused a rapid decline in the number of staff and the level of expertise possessed by the Air Force at its centers for research, development, acquisition, and maintenance. Nevertheless, the Air Force acquisition workforce has eagerly accepted major acquisition challenges. In some cases, the cost and schedule overruns incurred by these acquisition programs showed that the USAF was not adequately prepared and resourced to take on these challenges.[9] This lack of preparation has led to the creation of an increasing number of oversight panels and committees that demand the time and attention of the same workforce that is expected to execute already understaffed and overworked program organizations.

Subsequent directives from the Secretary of the Air Force and the Air Force chief of staff gave very specific instructions for upgrading and managing the acquisition workforce.[10] However, the combination of budget and hiring constraints; attrition through retirements; the competition from private industry and the speed at which they can hire technical and business graduates; the immediate need to retain an existing workforce to meet current demands; and the absence of confidence in acquiring stable and satisfying career advances and career development assignments—all appear to have had a negative impact on the cost, schedule, and performance issues documented in Air Force acquisition.

A recent report from the Defense Business Board[11] outlines an aggressive program to address the main factors that inhibit bringing highly qualified personnel into Pentagon leadership positions. After years of studies that repeatedly identify the same problems in the Air Force acquisition workforce, it is evident that more steps could be taken to restore acquisition excellence in the Air Force. The current incremental steps are not achieving the necessary results. The recommendations in this report identify some of the most important steps.

---

[8] Examples include the following: Office of the Assistant Secretary of the Air Force (Acquisition), *Acquisition Improvement Plan,* May 4, 2009, http://www.dodbuzz.com/wp-content/uploads/2009/05/acquisition-improvement-plan-4-may-09.pdf; G.E. Christle, D. Davis,and G. Porter, *CNA Independent Assessment: Air Force Acquisition,* Center for Naval Analyses, Alexandria, Va., 2009; G. Vernez and H.G. Massey, *The Acquisition Cost-Estimating Workforce: Census and Characteristics,* TR-708-AF, RAND Corporation, Santa Monica, Cal., 2009; J.A. Ausink, L.H. Baldwin, and C. Paul, 2004, *Air Force Procurement Workforce Transformation,* RAND Corporation, MG-214-AF Santa Monica, Calif.

[9] G.E. Christle, D. Davis, and G. Porter, *CNA Independent Assessment: Air Force Acquisition,* Center for Naval Analyses, Alexandria, Va., 2009, p. 3.

[10] Examples include the following: U.S. Air Force, Air Force Instruction 63-501, Air Force Acquisition Quality Program, May 31, 1994, Certified Current November 4, 2009, http://www.e-publishing.af.mil/; U.S. Air Force, Air Force Instruction 36-2835, Annual Acquisition Awards Programs, August 17, 2011, http://www.e-publishing.af.mil/; U.S. Air Force, Air Force Instruction 63-138, Acquisition of Services, May 21, 2013, http://www.e-publishing.af.mil/.

[11] Defense Business Board, *Selecting Senior Acquisition Officials,* April 21, 2016, http://dbb.defense.gov/Portals/35/Documents/Meetings/2016/20164/Sr%20Acquisition%20Officials%20Presentation%20-%20Approved%2021%20APR%202016.pdf, accessed August 17, 2016.

2

Strategic Value of Owning the Technical Baseline

Before discussing the barriers the Air Force will need to overcome to own the technical baseline in its acquisition programs, it is important to discuss the value to the Air Force of owning the technical baseline. Ownership of the technical baseline not only provides benefits but also helps to reduce risks. It is informative to view some of the benefits through the lens of examples within the Air Force as well as in other agencies that own the technical baseline for their programs. The following sections in this chapter will highlight the importance of owning the technical baseline, the risks associated with not owning it, and how similar agencies own the technical baseline of their programs.

THE IMPORTANCE OF OWNING THE TECHNICAL BASELINE

New weapons systems require enormous investments in systems and people, as well as in tactics, techniques, and procedures. The system itself needs to be well understood throughout its life cycle, from concept formulation through operational use. Modern warfighting systems operate within a system of systems, and the complex web of interfaces and integration nodes among these systems requires planning, investment, and support. The trade-offs and decisions required for optimal planning and execution of these advanced systems can only be made by the entity—in this case the USAF—responsible for the integration of all elements across the entire system and their eventual use to defend U.S. interests. In today's world, almost all of the systems the Air Force will acquire and integrate into its operations will be developed by the private sector and will demand a high degree of varied technical knowledge and understanding and communication between the acquisition and operational communities.

"Owning the technical baseline" means that program managers (PMs) and associated personnel have sufficient technical knowledge, experience, and authority to enable program success by making informed, timely, and independent decisions to manage cost, schedule, and performance risk, while ensuring disciplined and integrated program execution. In short, owning the technical baseline allows the government acquisition team to respond knowledgeably and effectively to systems development, operations, and execution and avoid technical and management barriers to mission success. In some cases the supplier has critical knowledge that the acquisition team does not possess—which can be exacerbated if the Air Force does not own the technical baseline. This can include not understanding the constraints that limit performance, which could result in the requirements not being met or can mean that those requirements are unrealistic for the technology readiness level (TRL) of the proposed technical solution. Owning the technical baseline makes certain that government personnel understand the chosen design, understand why that particular design and its various features have been selected over competing designs, and understand the alternative paths to the final product in the face of unanticipated cost, schedule, and performance challenges.

To be effective, the entire government acquisition team, not just the PM or the chief engineer, needs to collectively own the technical baseline; PMs, contracting officers (COs), engineers, budget managers, and maintainers make or contribute to decisions that depend on their knowledge of the technical baseline, and all share a common responsibility for mission success. The entire government acquisition team needs to be able to understand the implications of a proposed change, both its current effects and its downstream effects, on a system's functional and sustainable capabilities; on program risk, schedule, and cost; and on contractual details. Informed decision making, particularly when dealing with large, complex, and technically advanced weapon systems, depends critically on the ability of the government acquisition team to control, understand, and modify the technical baseline when necessary.

THE RISK OF NOT OWNING THE TECHNICAL BASELINE

Loss of the technical baseline has many ramifications. First and foremost, the government acquisition team needs to understand the technical and programmatic risks, as well as mitigating strategies to limit their impacts, associated with developing and fielding the required capability. If the team relies solely on a prime contractor for this assessment, the government's decision-making capability is supplanted. As a result, the time lines may lengthen because the team needs more time to work through technical challenges it may not fully understand. More government resources, in terms of both people and money, may be required to complete the work, threatening the overall budget, the program objectives, and the Air Force missions themselves. The risk of late delivery of operational capabilities begins to rise, with potential negative impacts on the interoperability of system elements and on the ultimate integration of the product or service into either a new system or the overall system of systems. System support over the life cycle is likely to be more expensive, while its operational capability may fall short of need and expectations. The entire program investment may be imperiled by poorer performance, thus failing to meet the standards required by the operators, program offices, and Air Force leadership. The benefit of owning and the risk of not owning the technical baseline can be illustrated by several examples contained in Box 2.1.[1]

HOW OTHER AGENCIES OWN THE TECHNICAL BASELINE

Owning the technical baseline is essential to assuring that the technology the USAF is acquiring meets its operational needs. Put differently, the Air Force performs missions that are uniquely military and inherently governmental. The Air Force values operational input when acquiring technology that is essential to mission performance. Accordingly, for the most part and as in the other Services, the PMs who occupy key leadership positions in the Air Force acquisitions arena are mostly uniformed officers supported by a largely civilian technical workforce.

There are numerous examples in other federal agencies and departments in which ownership of the technical baseline of programs and existing systems is achieved.[2] Such agencies and departments typically place a high value on the technical proficiency of their workforce and view the operation of the government acquisition team as a necessary complementary unit. These teams most often include experienced technically grounded program managers who possess a strong sense of mission, responsibility, and accountability and who are supported by a well-trained engineering and technical cadre, including experienced contracting, financial, and other functional personnel who are considered and treated as an integral part of the government acquisition team.

[1] The examples in Box 2.1 are representative vignettes and do not contain all background and contextual information. Information for each example was provided by Air Force Life Cycle Management Center representatives. While they are documented in an abbreviated manner in this report, the testimony heard by the committee demonstrated more fully the consequences of abdicating decision making, reclaiming it, and owning it from the start.

[2] Representatives were interviewed from the following agencies and departments: the Department of the Navy, the Department of Defense, the National Aeronautics and Space Administration, the Missile Defense Agency, the National Nuclear Security Administration, Defense Acquisition University, and Defense Innovation Unit-Experimental. A full list of meeting participants can be found in Appendix C.

BOX 2.1
Technical Baseline Examples—
Benefits of Owning and Risks of Not Owning

The A-10

While the Air Force owned the technical baseline for the majority of its programs prior to the 1990s, the A-10 highlights a case during that time period when the Air Force did not own the technical baseline and was forced to reestablish ownership. Regaining the lost technical baseline proved to be an enormous task and took several years to complete, resulting in delays to several crucial updates. For most of the A-10's operational lifetime, the prime contractor, Fairchild Industries, owned the technical baseline. In the late 1990s, the company was acquired and only in 2003 did the Air Force assume ownership of the technical baseline. This required an investment of approximately $7 million for tooling drawings and incorporation of outstanding USAF-generated production drawing changes. Lockheed Martin continued through 2011 as a prime integrator. Current annual costs for tools and software required to manage the technical baseline in structures is $400,000. The Systems Program Office (SPO) spends around $23 million annually on systems engineering tasks. The SPO has found several benefits of owning the technical baseline, including increased capability to rapidly evaluate and mitigate problems found in the field and a greater ability to evaluate the technical content for major purchases. Challenges remain, primarily in technical agility and lack of funding for training the organic workforce. Original equipment manufacturers or shared resources still need to be developed and maintained for low-use, high-value skill sets. For sustainment there will be a need for specialty engineering in tooling, materials, and process engineering to coordinate requirements between the SPO, the supply chain, and the depot organizations.

The Launch Test Range System

The Launch Test Range System program is responsible for safe tracking and positive control of vehicles in both the Atlantic and Pacific test ranges. Driven by concern over the erosion of technical competence and capability, the PM decided to create a new systems engineering and integration contract, with 50 full-time equivalent positions. The current team structure consists of 59 SE&I contractors and 30 organic engineers—a mix of civilian, military, and federally funded research and development center subject-matter experts. The annual cost of this team is roughly $20 million. The sum of the various actions taken resulted in estimated savings of

Top leadership support within each department or agency, for both the missions and the programs, is an essential part of successful ownership of the technical baseline. Successful ownership of the technical baseline in other departments and agencies allows the leadership to make key and other integral decisions in program planning, budgeting, and execution with participation and buy-in across the entire organization. Those agencies and departments maintain an acquisition workforce that is staffed and balanced appropriately, and they clearly spell out and document PM roles, responsibilities, accountabilities, and authorities. Moreover, they view the acquisition force as a vital functionary and mission enabler, with the ability to understand and execute the complexity of acquiring new systems.

$700 million over the 10-year contract. The program office can now make informed award decisions and handle prioritization and even predictive sustainment—an example of a benefit of owning the technical baseline.

The F-16

F-16 modification integration and hardware engineering are performed in-house, and this program demonstrates the value of owning the technical baseline. The F-16 SPO owns the technical baseline and the government team executes program management. In addition, there is an integrated government/contractor team to provide engineering support and configuration management. An example of how well this works is the emphasis on competitive source selections for equipment and upgrades, which can include development of engineering drawings, tech orders, installation and integration, and testing. The move to in-house maintenance for five systems alone on the F-16 saves the Air Force over 50 percent in recurring costs.

The KC-46

The new Air Force tanker, the KC-46, is a commercial derivative (Boeing 767). However, it is really a hybrid—a commercial airliner with some military-unique systems and missions. When consideration is being given to owning the technical baseline on a hybrid platform, it is important to understand the technical pedigree of specific systems and subsystems. For example, a key component is the boom. This system contains sophisticated and unique automatic control and stabilization augmentation features. As of May 2016, the certification of the boom has not met schedule, and delivery of the aircraft is projected to be at least a year delayed.[1] The development contract is fixed price, but that does not mean that the original equipment manufacturer bears all the costs—recently four airborne refueling tests had to be aborted with costs to the Air Force including fuel, labor, and aircrew/aircraft availability. According to leadership at the Air Force Life Cycle Management Center the boom had not been identified as a critical technology element, something that would most likely have been identified as such had the program office properly owned the technical baseline. Such identification would necessitate a control systems expert from the program team assessing the readiness of the technology throughout its development. Exacerbating this was an uncertainty on the part of the government over data rights.

[1] L. Seligman, "Boeing's KC-46 Tanker Will Miss Major Deadline," *Defense News*, May 27, 2016.

As seen through examples in other agencies, successful government acquisition teams possess clear leadership support that openly promotes a risk-tolerant culture and demonstrates a clear sense of urgency to achieve the organization's goals. PMs have unquestioned accountability for the overall program, which in turn provides a compelling motivation for them to own their technical baseline. A professionally managed and valued engineering workforce with defined career paths and reward structures is essential to producing capable PMs, as is having a program management workforce that has prolonged and relevant experience with industry. A well-managed program management workforce is characterized by PMs who possess continuity and longevity in their positions and a wealth of experience within their domains, matched with the necessary continuing education and training as well as mentoring opportunities both up and down the chain of command.

The National Nuclear Security Administration (NNSA), in addition to maintaining the nation's nuclear stockpile, builds nuclear propulsion systems that are typically first of a kind. Similar to many Air Force programs, safety and long-term, high-quality performance are critical concerns for these systems. NNSA owns its technical baseline in part by using the Department of Energy's national laboratories as a deep technical bench for subject matter expertise. Programs use clearly defined technical requirements to guide an integrated team of program officers, contractors, and end-users toward a set of documented program objectives.

NNSA has also developed a federal risk register to perform its own technical risk assessments, which are subsequently compared to and reconciled with periodic independent external assessments. The risk management guidelines are part of a set of guidelines for project and program management that follow industry best practices and have been tailored to the NNSA environment. This practice fosters an awareness of a program's baseline and helps prevent small problems from becoming larger ones. Additionally, NNSA enforces management practices that work to retain talented practitioners, and to cultivate a sense of mission and dedication to the program. All of these attributes are considered vital for owning the technical baseline.

The U.S. Navy, like the USAF, acquires specialty military technology that meets demanding operational requirements, many of which are unique to the naval service. The Navy has chosen a different and more nuanced approach to reconciling the need for a distinctly service-oriented operational point of view by providing education, training, experience, and succession planning to enable its personnel to perform competently in the operation of technology acquisition. The Navy employs a tiered approach for developing personnel to effectively lead acquisition programs. There is a cadre of officers who enter the Navy's engineering workforce after a few years of operational experience as junior officers and postgraduate education in an engineering discipline. The careers of these officers, designated as Engineering Duty Officers or Aeronautical Engineering Duty Officers, are managed so that some specialize in the maintenance of systems and some specialize in acquisition development programs. These officers are restricted to serving in engineering positions within the Navy. PMs for major programs are often selected from this cadre of technically educated, experienced engineering duty officers. Typically these officers have had several assignments supporting program managers in activities such as test and evaluation, research and development, or the administration of acquisition contracts in the field before they are assigned as program managers for major acquisition programs. In general the career paths for engineering duty officers are managed by repeat tours in one of the warfare specialty subsets of the Navy (i.e., submarines, Aegis surface ships, aircraft carriers, aircraft, etc.) all the way to flag officer rank.

The Navy also trains and educates an "unrestricted" cadre of officers who specialize in acquisition-related activities during shore duty and may later enter the ranks of materiel professionals. These officers qualify as acquisition materiel professionals[3] through obligatory training. This training and experience allows these officers to bring more recent operational experience to the government acquisition team. Senior military leaders for Navy acquisition programs are selected from the ranks of engineering duty officers, aeronautical engineering duty officers, or materiel professionals, whose backgrounds emphasize and focus on the importance of technical and engineering experience.

Owning the technical baseline provides the natural benefit of a rewarding acquisition career path that allows assimilation of crucial operational knowledge. The committee considered distinctions between the Navy and Air Force training, and discussed whether the Air Force should adopt the Navy's engineering duty officer career path as a model for improving the technical acumen and operational experience of acquisition personnel. This latter approach evoked mixed reviews by several current Air Force program managers when it was raised in the interview process. Generally, concerns over implementing the engineering duty officer career path in the Air Force were focused on assuring a high level of operational experience and knowledge, the potential disruption of current promotion processes, and the ability to gain broader experience in the Air Force. These concerns when the Air Force was considering an alternative strategy for developing a more professional cadre of experienced Air Force officer acquisition personnel did not seem to be based on an analysis of alternative approaches. Operational experience and concerns are also critically important in Navy acquisition programs. The Navy officer professional acquisition cadre remain keenly aware of operational needs throughout their careers. A contributing element in the way senior PMs are selected in the Air Force may stem from a fundamental career development problem in that many young people are needed to operate Air Force aircraft and systems, but few are needed to fill senior leadership positions in operations. A way of addressing this personnel management problem has been to "lateralize" capable Airmen into senior positions in the acquisition field despite little or no acquisition experience.

As mentioned previously, there are numerous examples in other federal agencies and departments in which ownership of the technical baseline of programs and existing systems is achieved; this report highlights only a few of those examples.

[3] The Defense Acquisition University defines "materiel management" as "direction and control of those aspects of logistics that deal with materiel, including the functions of identification, cataloging, standardization, requirements determination, procurement, inspection, quality control (QC), packaging, storage, distribution, disposal, maintenance, mobilization planning, industrial readiness planning, and item management classification. Encompasses materiel control, inventory control, inventory management, and supply management."

Representatives from other agencies interviewed by the committee regarded ownership of the technical baseline as an essential element of successful acquisition, and they managed their acquisition workforce to emphasize acquisition experience in those individuals responsible for managing programs. The details of exactly how that experience is acquired vary among the agencies. The Air Force cannot implement all of the methods for owning the technical baseline used by other agencies and departments, but it can recognize that policy, education, and experience all need to emphasize the importance of owning the technical baseline for Air Force programs.

The next chapter discusses barriers to owning the technical baseline identified within the Air Force and the subsequent recommendations for overcoming those barriers. These recommendations were developed based on the information presented to the committee by the Air Force as well as by other agencies and departments.

3

Recommendations and Barriers to Implementation

It is beneficial to address the barriers the Air Force faces in owning the technical baseline of its acquisition programs in the context of the key topics outlined in the report of the workshop on ownership of the technical baseline (OTB) (leadership and culture, workforce management, contracting support, and funding). Throughout the following sections, the committee will highlight the key barriers to the Air Force owning the technical baseline of its acquisition programs and will offer recommendations for eliminating them.

LEADERSHIP AND CULTURE

It is clear that a strong sense of shared mission ownership, declared and upheld by senior leadership and shared down the chain of command, is critical to fostering quality program management and successful program outcomes. This section discusses leadership and cultural barriers to owning the technical baseline in Air Force acquisition programs. This discussion highlights the importance of consistent tenancy in the Air Force's acquisition leadership positions and the need for changes in the Air Force's risk-averse culture.

The Air Force has ceded ownership of the technical baseline in many acquisition programs. The process began with decisions made and executed by the Department of Defense (DoD) and Air Force leadership in the mid-1990s and evolved over time. The ongoing tension between funding for operational priorities and support for acquisition personnel; indiscriminant application of the total systems performance responsibility (TSPR) approach; ethics violations involving Air Force senior acquisition personnel in the acquisition community; and an erosion of the organic technical workforce in response to budget reductions have engendered a series of actions and reactions that have negatively impacted the Air Force's organizational culture. This culture—defined as a system of shared assumptions, values, and beliefs that governs and influences how people act, interact, and perform their jobs—is instrumental in the Air Force's ability to own the technical baselines of its current and future programs. In addition, there has been a clear shift in recent years toward a risk-averse culture within the Air Force acquisition community. Over the past two decades the Air Force has adopted a culture of "just saying no," replacing the team-oriented culture of "here is a way to achieve the technical mission and objectives within our legal and ethical bounds."[1]

A key issue in the Air Force acquisition community is a persistent lack of continuity in the position of Assistant Secretary of the Air Force for Acquisition (SAF/AQ), especially in relation to the continuity seen in other senior Air Force positions. Figure 3.1, compares the amount of time several president-appointed, Senate-confirmed USAF positions were designated as "acting" or "vacant" since January 2000.[2] The figure clearly shows the disparity between the SAF/AQ position and the other Senate-confirmed positions in the Air Force. It is unclear why this disparity exists. While the Air Force does not control all aspects of the nomination and confirmation processes, it can still advocate that the SAF/AQ position be filled as quickly as possible within the constraints and timelines of the process. Prolonged vacancies of the SAF/AQ position have, over time, eroded the necessary senior leadership and hierarchical support for program executive officers (PEOs) and program managers (PMs), particularly when making potentially controversial decisions about mission-critical defense programs.

CONCLUSION 1: Consistent tenancy in the position of Assistant Secretary of the Air Force for Acquisition will help to revitalize, focus, and provide visible support for the acquisition community's critical role in program development and execution.

RECOMMENDATION 1: The Secretary of the Air Force should investigate why the position of Assistant Secretary of the Air Force for Acquisition is in an acting or vacant status more frequently than other Air Force Assistant Secretary positions. This investigation should consider how the Air Force, along with other Services and government agencies, fills similar critical positions and

---

[1] Vice Admiral David A. Dunaway (USN, retired), former commander of Naval Air Systems Command, interview with the committee on February 8, 2016.

[2] U.S. Air Force, *Key Personnel*, Headquarters United States Air Force, Air Force Historical Studies Office, January 2013, http://www.afhso.af.mil/shared/media/document/AFD-130410-035.pdf.

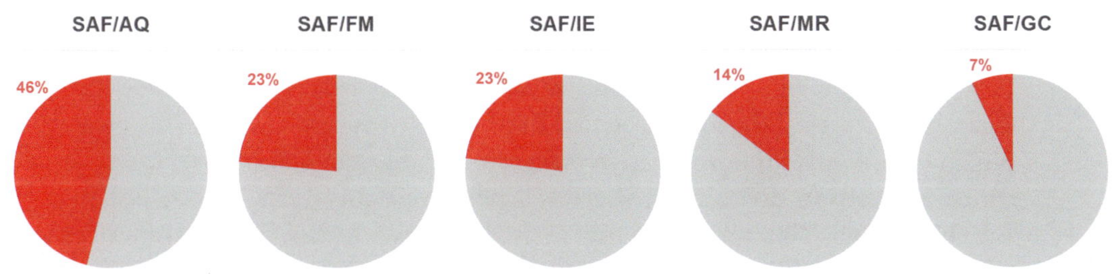

FIGURE 3.1 Percentage of time (in red) from January 2000 to January 2016 that Air Force Assistant secretary positions, including the General Counsel (SAF/GC), were designated as either "acting" or" vacant." Positions included SAF/AQ, Assistant Secretary of the Air Force for Acquisition; SAF/FM, Assistant Secretary of the Air Force for Financial Management and Comptroller; SAF/IE, Assistant Secretary of the Air Force for Installations, Environment and Logistics; SAF/MR, Assistant Secretary of the Air Force for Manpower and Reserve Affairs. Figure does not include the positions Secretary and Under Secretary of the Air Force.

should focus on identifying best practices for implementation. The Assistant Secretary of the Air Force for Acquisition position should not be vacant for any extended period of time, and the use of an acting individual should be minimized. Furthermore, in order to attract competitive talent, the Air Force should ensure that it does not impose any additional restrictions beyond those required by law, especially relative to the post-employment period, for the position of Assistant Secretary of the Air Force for Acquisition.

The Air Force acquisition culture emphasizes process and the pursuit of perceived cost reductions. The risk-averse culture of Air Force acquisition is governed primarily by process compliance, the cost of which is estimated to account for nearly 25 percent of every dollar spent.[3] The high level of oversight[4] in place

---

[3] The Honorable Stan Soloway, President and CEO, Professional Services Council, interview with the committee on January 14, 2016.

[4] The Under Secretary of Defense, the Honorable Frank Kendall said in his memorandum "Implementation Directive for Better Buying Power 3.0—Achieving Dominant Capabilities through Technical Excellence and Innovation," section "Streamline documentation and staff review" the following: "In [Better Buying Power] BBP 2.0, we tracked how much time is logged to prepare for staffed document reviews and decision review briefings. The Government Accountability Office has also recently released a study on document lead times and value. Our data indicates that excessive program management time is spent supporting staff reviews and preparing documents primarily for review, instead of focusing on program execution. The Department will continue and increase the effort to reduce documentation and reviews. Program managers are expected to suggest tailoring throughout the program lifecycle" (April 9, 2015, p. 20, http://www.acq.osd.mil/fo/docs/betterBuyingPower3.0(9Apr15).pdf).

for military programs, and the weight of meeting the oversight requirements, constrain and hinder programs and program staff.[5] A symptom of a risk-averse culture is that personnel often use the most restrictive interpretation of a policy, and even though processes allow for the elevation of issues, it is rarely done. This risk-averse posture can hinder the innovative problem-solving mindset of even the most seasoned acquisition executive.[6] There is a perceived culture of acceptance in some Air Force acquisition programs that fosters the development of program managers who only verify the existence of results from the contractor and perform minor, if any, independent technical verification and validation. These program managers, along with their acquisition teams, were forced to abandon their role of organic engineering analysis by policy constraints and funding reductions. However, such organic engineering analysis is often necessary to assure that program and technical decisions best meet mission requirements. This new paradigm threatens the ability of the Air Force acquisition enterprise to deliver "war-winning" capabilities within cost and on schedule. Oversight is replacing program risk management and is actually creating more program risk by reducing verification and validation.

In an Air Force culture that has devalued the role of acquisition management and personnel, program management positions have often come to be regarded as career path stepping-stones rather than as coveted and important leadership positions. Program managers and engineers—those who have the most knowledge of program risk—have a seat at the table but often do not have respected input when making decisions related to program risk and are largely unappreciated in terms of both numbers and authority for technical and program management. Engineers in the Air Force have a voice, but they often do not have a vote.[7] These practices are in contrast to the acquisition era prior to TSPR, when engineers not only had a voice and a vote but often were the prime assessors of program technical risk and the associated cost, schedule, and performance risk.

CONCLUSION 2: The current risk-averse culture, along with the gap in technical engineering expertise within Air Force acquisition programs, hinders program managers from making informed, timely, and independent decisions. This culture is negatively impacting programs and is a driver of rising costs and protracted schedules.

[5] Blaise Durante, Director, Blaise J. Durante & Associates, Inc., interview with the committee on February 9, 2016.

[6] The Honorable Jack Gansler, professor emeritus, University of Maryland School of Public Policy, and Former Undersecretary of Defense for Acquisition, Technology, and Logistics, interview with the committee, February 8, 2016.

[7] Jorge Gonzalez, Director of Engineering and Technical Management/Services Directorate, Air Force Life Cycle Management Command (AFLCMC), interview with the committee, March 30, 2016.

RECOMMENDATION 2: Air Force senior leaders should define, develop, and execute a strategy that balances risk and reward from a program implementation viewpoint, fosters a learning environment characterized by healthy tension and debate, and actively rewards acquisition personnel that regularly find a "pathway to yes." A risk-tolerant[8] acquisition culture, in concert with a sense of urgency, is critical to agile and timely acquisition for the Air Force to maintain its advantage against rapidly evolving threats. Significant attention should be given to the proliferation and acceptance of this crucial change. The strategy should include the following at a minimum:

- Establishing an education and training program to promote and develop a risk-tolerant culture that includes the use of current and former experienced acquisition professionals to provide guidance and mentorship.[9]
- Encouraging the pursuit of more reasonable interpretations of policy and process flexibility to more efficiently accomplish program goals while maintaining compliance.
- Assuring that logical and reasonable deviations from policy or requirements can be expeditiously pursued by empowered acquisition personnel.

WORKFORCE MANAGEMENT

The following discussion of workforce management addresses the need to adequately staff technical positions within the Air Force, as well as the need for consistent and continuous management in key programmatic roles, sound career management practices to retain engineering and acquisition talent, and the use of appropriate contracting vehicles to support the technical workforce where necessary. In addition, widely recognized best practices in both the federal agencies and industry are available for the Air Force to consider and employ.

The Air Force has gradually reduced its organic technical workforce[10] through a combination of service downsizing, devaluing technically trained personnel, cost-

[8] Paraphrasing from David Hillson's book *Effective Opportunity Management for Projects, Exploiting Positive Risk* (Marcel Dekker, New York, 2004), the term "risk tolerant" is viewed as "being reasonably comfortable with most uncertainty and accepting it exists as a feature of life or business." "Risk averse," on the other hand, is viewed as "being uncomfortable with uncertainty, having a low tolerance for ambiguity, and seeking security and resolution in life and business."

[9] This recommendation aligns with points outlined in the 2013 memorandum of the Under Secretary of Defense for Acquisition, Technology, and Logistics on key leadership positions and qualification criteria (November 8, available at https://acc.dau.mil/adl/en-US/684463/file/75211/USA001464-13%20USD(AT_L)%20Key%20Leadership%20Positions%20and%20Qualification%20Criteria%20Memo%20(8%20Nov%2013).pdf).

[10] Organic technical workforce includes both military and government civilians.

cutting measures, and attrition. Additionally, program growth has occurred simultaneously with this loss of workforce. This gradual loss, as evident in Figure 3.2, in some cases led the Air Force to assign personnel who lack the necessary technical education or expertise to the role of PEO or PM, handing over control of the technical baseline to prime defense contractors.[11] Accordingly, the duties of technically trained Air Force personnel have evolved—once directly engaged in technical work, they now primarily monitor contractors doing that work.

The Under Secretary of Defense's April 2015 memorandum to the Services on "Implementation Directive for Better Buying Power 3.0—Achieving Dominant Capabilities through Technical Excellence and Innovation," provides guidance to strengthen organic engineering capabilities. The memorandum specifically states that[12]

> DoD cannot effectively support the Warfighter nor retain its technological superiority without a competent and innovative organic engineering workforce, both military and civilian. The goal of this initiative is to strengthen our organic engineering capabilities by equipping our technical workforce with essential education, training, and job experiences, along with the right physics-based tools, models, data and engineering facilities to efficiently and effectively manage the technical content of our complex products throughout their lifecycle. The Department also needs to take active steps to strengthen organic engineering capabilities to better understand the technical risks associated with program execution for its development programs, and this requires a strong engineering workforce.

It is clear that, at present, the Air Force cannot own the technical baseline in all appropriate programs. The ability to own or regain ownership of the technical baseline in the Air Force is complicated by a lack of capacity to meet current and emerging engineering and technical demands. This is driven in part by the lack of a clear and valued career path for uniformed engineers in the Air Force, which hinders development of adequately qualified acquisition personnel.[13] Careful analysis will be needed for the Air Force to determine how it can create career paths for uniformed engineers so there will be enough senior, technically competent, and programmatically experienced acquisition professionals to manage major acquisition programs. An effective analysis will specifically look at the way the Navy and other services and agencies manage their technical resources. The Air Force's situ-

[11] Douglas L. Loverro, Deputy Assistant Secretary of Defense for Space Policy, Department of Defense, interview with the committee on March 30, 2016.

[12] Under Secretary of Defense, "Implementation Directive for Better Buying Power 3.0, Achieving Dominant Capabilities through Technical Excellence and Innovation," April 9, 2015, p. 23, http://www.acq.osd.mil/fo/docs/betterBuyingPower3.0(9Apr15).pdf.

[13] Jorge Gonzalez, Director of Engineering and Technical Management/Services Directorate, AFLCMC, interview with the committee on March 30, 2016.

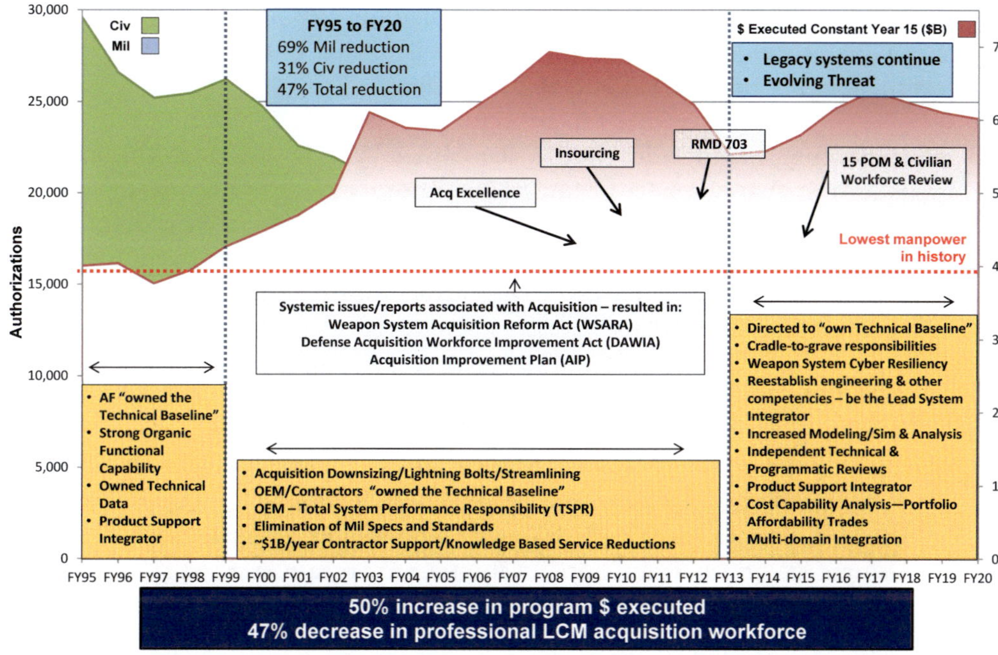

FIGURE 3.2 Air Force Life Cycle Management Center (AFLCMC) Acquisition Workforce. NOTE: AF, Air Force; FY, fiscal year; OEM, original equipment manufacturer; POM, Program Objective Memorandum; RMD, Resource Management Directive. SOURCE: Lynn Eviston, Director, Plans and Programs, Air Force Life Cycle Management Center. Approved for release by SAF/PA on February 8, 2016, Case Number 2016-0505.

ation is further complicated by the fact that there are simply not enough engineers within the Air Force Life Cycle Management Center (AFLCMC) to meet current demands[14] for technical work, let alone to take on additional programs.[15]

The authorizations in fiscal year (FY) 2020 for the acquisition workforce in the AFLCMC are projected to result in an overall reduction of 47 percent below FY1995 levels.[16] Additionally, the ratio of program engineers to program managers has dropped to 4:1, as contrasted to the previously robust high of 10:1.[17] It is difficult to qualitatively assess what effect such a reduction in engineering support

[14] Appendix D consists of a memorandum on owning the technical baseline by Lt. Gen. John F. Thompson, the Commander of the AFLCMC, dated June 20, 2016.

[15] Ibid.

[16] See Figure 3.2.

[17] Jorge Gonzalez, Director of Engineering and Technical Management/Services Directorate, AFLCMC, interview with the committee on March 30, 2016.

has had, but it is notable. Experience shows that the responsiveness and timeliness of the engineering staff to technical matters is related to the depth of experience and the number of qualified engineers available.

The USAF has found itself in an acquisition crisis. Whenever it needs to make informed, timely, and independent decisions related to the technical baseline it needs to own that baseline. Owing to the resource constraints and the fact that acquisition programs in the Air Force are in different stages of their development; have varying contract mechanisms and end products; and have various ranges of technical complexity, there is no "silver bullet" for fixing the issue of owning the technical baseline.

CONCLUSION 3: The USAF is "over-programmed," and its organic technical workforce is critically understaffed. This combination is highly detrimental both to the sustainability of current weapon system programs and to the health and success of future programs. The reduction in the organic workforce, coupled with a loss of technical education and experience, has subsequently hampered the Air Force's ability to attain or regain control of the technical baseline when it is most needed.

RECOMMENDATION 3: The Air Force should continue and complete its efforts to determine which current programs should own the technical baseline and develop staffing standards to determine the proper mix and number of military and civilian engineers required to own the technical baseline for those programs.[18] Criteria should be established for when the Air Force should own the technical baseline[19] as opposed to having knowledge of the baseline as technical integrator or interface systems reviewer. The decision to own the technical baseline for future programs should be included in the acquisition milestone protocol as gated decision points. Additionally, the Air Force should develop methods to measure whether or not selected programs have successfully achieved, and are maintaining, ownership of the technical baseline. Cost overruns, schedule delays, and unidentified, or incorrectly identified, key performance parameters (KPPs) are potential measurement points.

---

[18] This recommendation is in line with Recommendation 2-2 from the National Research Council report *Examination of the U.S. Air Force's Science, Technology, Engineering, and Mathematics (STEM) Workforce Needs in the Future and Its Strategy to Meet Those Needs.* That report says "the Air Force should review and revise as appropriate its current requirements and preferences for personnel with STEM capabilities in every career field and occupational series" (The National Academies Press, Washington, D.C., 2010, p. 32).

[19] Examples of programs for which the Air Force could apply and test the developed criteria would be, but are not limited to, B-21, GPS OCX, Joint Stars Recapitalization, DCGS, GPS III.

In the acquisition community, the personnel rules and regulations that govern the assignment type, variety, and tour length necessary for promotions in rank and career advancement have had a negative impact on program management. Civilian and military personnel typically rotate out of programs too rapidly to acquire the experience, insight, competence, and confidence necessary for managing large, complex, and indispensable defense programs.

These short tours hinder or are detrimental to the management of the technical workforce, which in turn prevents effective ownership of the technical baseline and can negatively impact cost and schedule. Even the most capable leaders grapple with a demanding learning curve when they first enter a new position, while the workforce simultaneously adjusts to the new leadership's priorities, needs, and intents. This further hinders the program's progress, effective decision making, and success within a timeline that does not compensate for personnel turnover and often does not account for a transition period from one program manager to the next. Major programs demand continuous attention to oversight, as well as attention to emergent technical, business, and funding issues in order to be effective and successful.

Historically, the USAF valued the military members of the engineering workforce and provided retention bonuses to them for up to 10 years so the USAF could retain them for an entire 20-year period.[20] The funding for these types of incentives appears to have been realigned to "higher priority" areas of the Air Force or totally zeroed out. However, the first overrun on a major defense acquisition program (MDAP) would more than pay for engineering bonuses[21] and the lost incentives that were so crucial to delivering successful programs in the past. A small portion of the approximate $14 billion cost overrun of the Air Force's space-based infrared system (SBIRS) could easily fund the essential growth in numbers and retention of engineers and technical personnel necessary to regaining technical capabilities in the USAF workforce.[22]

CONCLUSION 4: The Air Force currently lacks personnel stability, driven by personnel rotation and lengths of assignments, in its program offices, thus impacting program knowledge management within the program office.

RECOMMENDATION 4: The Air Force should review, and make appropriate changes to, current assignment policies and practices for the acquisition workforce to reduce turnover and attrition and increase succession and transi-

---

[20] Douglas L. Loverro, Deputy Assistant Secretary of Defense for Space Policy, interview with the committee on March 30, 2016.

[21] Ibid.

[22] M. Gruss, Unlocking the SBIRS Data Revolution, *Space News Magazine*, April 25, 2016, pp. 11-13 and 25, http://www.spacenewsmag.com/feature/unlocking-the-sbirs-data-revolution/.

tion planning; should invest in a more structured mentoring program across the acquisition workforce to increase the sharing of best practices; and should ensure that the career management system for the acquisition workforce be charged with providing appropriate educational opportunities, training, and industrial experiences to acquisition personnel. The intent of the review should be to create strong career paths for acquisition personnel, reflecting the critical value of acquisition to future Air Force operations.

Owning the technical baseline is both critical to, and relies on, effective program management with technically competent individuals who can make informed and timely decisions that are critical to mission success. *Better Buying Power 3.0, Interim Release* noted as follows:

> We would not expect to see a non-lawyer supervising a group of trial lawyers litigating cases, and we would not expect to see a non-surgeon supervising a group of doctors performing surgery. We should also not expect a Program Manager with no technical education or experience in engineering to supervise a development program.[23]

The Air Force Officer Classification Directory[24] (AFOCD) describes the Air Force's Acquisition Utilization Field[25] as encompassing "staff and management functions peculiar to the Air Force acquisition life cycle." The directory continues as follows:

> It is desirable that entry into the career field be preceded by assignment in another utilization field whenever possible. Officers who enter the career field on their initial tour should seek a subsequent assignment in another utilization field followed by a return to the acquisition program management career field. This desired career broadening is to provide a better perspective and understanding of the interfaces between functions of acquisition management and related functions in the developing, operating, training, and support commands. Lateral inputs will include only those officers who have clearly demonstrated a potential for effective administration and program management beyond their basic specialty.

In FY1995,[26] there were more than 6,500 Air Force officers in the Developmental Engineering Utilization Field (62XX) and the Acquisition Utilization Field (63XX). Of those officers, 36 percent possessed a bachelor's degree and 60 percent

---

[23] F. Kendall, *Better Buying Power 3.0, Interim Release*, September 19, 2014, p. 10, http://www.acq.osd.mil/dpap/sa/Policies/docs/BBP_3_0_InterimReleaseMaterials.pdf.

[24] U.S. Air Force, *Air Force Officer Classification Directory (AFOCD), The Official Guide to the Air Force Officer Classification Codes*, 2013, p. 216, http://www.uc.edu/content/dam/uc/afrotc/docs/UpdatedDocs2013/AFOCD_30Apr13.pdf.

[25] USAF Specialty Code 63XX.

[26] Data were filtered by FY1995 and FY2015, active duty Air Force officers only, Air Force Specialty Codes 62XX and 63XX only, and education level (highest).

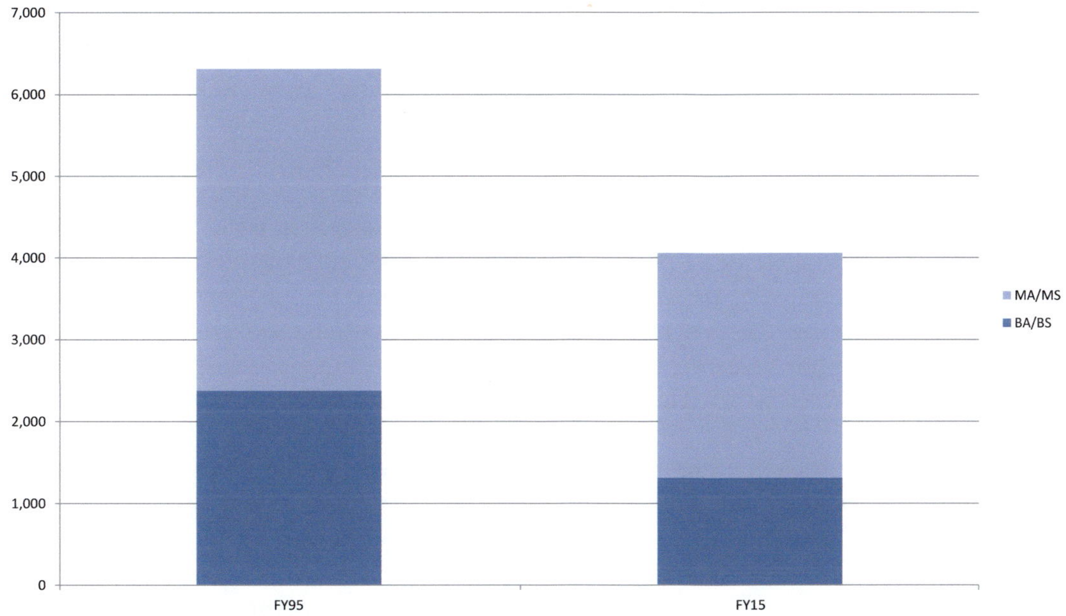

FIGURE 3.3 Highest education levels reported by Air Force officers classified as 62XX and 63XX.

possessed a master's degree as their highest education level. In FY2015, the total number of officers in the 62XX and 63XX fields had dropped to just over 4,300. Of those officers in FY2015, 30 percent possessed a bachelor's degree and 64 percent possessed a master's degree as their highest education level.[27] Figure 3.3 highlights the sharp decline in the total number of officers in the 62XX and 63XX between FY1995 and FY2015.

While the data presented above confirm the previously discussed decline in the number of Air Force 62XX and 63XX officers, they pertain only to the education levels of those officers and do not account for past technical experience. The committee heard anecdotal evidence that, until the 1980s, the desire for assignments in other utilization career fields prior to entry into the Acquisition Utilization Field (63XX) was regularly adhered to. Today's Air Force acquisition leaders and personnel system appear to often allow a new entrant to become a 63XX without adhering to the prerequisite of technical experience.[28] Over the same time span, other

[27] Data extracted from Air Force Personnel Center, "Report Builder Step 1 of 3," accessed August 17, 2016, http://access.afpc.af.mil/vbinDMZ/broker.exe?_program=ideaspub.IDEAS_Step1.sas&_service=pZ1pub1&_debug=0.

[28] Douglas L. Loverro, Deputy Assistant Secretary of Defense for Space Policy, interview with the committee on March 30, 2016.

organizations, such as the National Reconnaissance Office (NRO), have required program managers to demonstrate technical know-how and experience. Additionally, the USAF used to fund Reserve Officers' Training Corps (ROTC) engineering scholarships and would assign as many as 70 USAF engineers to ROTC as trainers and mentors to groom future engineers.[29] These ROTC trainers received bonuses to take on this valued and critical assignment.

Academic education levels and experience in other assignments prior to entering acquisition career fields fail to account for experience outside of the Air Force. The Education with Industry (EWI) program dates back to the birth of the Air Force in 1947.[30] The program is, as per the program handbook, "a highly selective, competitive non-degree educational assignment within an industry related to the student's career field." The EWI program was originally developed because Air Force leadership "determined that it needed a corps of talented officers who were capable of understanding the inner workings of the defense industry."[31] The need for both military and civilian personnel to have experience with industry—which gives them a better understanding and appreciation of how technical decisions are made in industry, industry incentives, contract terms and conditions, and contract incentives—is an essential part of the Air Force's ability to own the technical baseline.

> **CONCLUSION 5:** Successful program managers have commonly held the following qualifications and career attributes: a technical degree in a STEM field, operational assignments, education in business management, experience in either a business setting or the Education with Industry program and a transition into an acquisition role no later than mid-career.[32]

[29] Ibid.

[30] Air Force Institute of Technology, *Education with Industry Handbook*, August 2009, p. 1, https://www.afit.edu/cip/docs/EWI_Handbook.pdf.

[31] Ibid., p. 1.

[32] The memorandum of the Under Secretary of defense for Acquisition, Technology, and Logistics on key leadership positions and qualification criteria, dated November 8, 2013, stated that five factors had been identified as requirements essential for selection of key leadership positions for critical acquisition functions. The five requirements identified in the memorandum were education, experience, cross-functional competencies (executive leadership, program execution, technical management, and business management), tenure, and currency.

RECOMMENDATION 5: Air Force leadership should, in concert with its current activities,[33] ensure that there is necessary guidance and governance for the currency of appropriate skills of the acquisition workforce at all levels. This must include, but is not limited to, emphasis on the criticality to program success of technically educated and technically experienced program managers. Additionally, the Air Force should prioritize education and experience in industry, recognize its importance to the development of competent acquisition personnel, and increase the opportunities for members of the acquisition workforce to gain this education and experience.

RECOMMENDATION 6: The Air Force should establish, select, and equip a dedicated line of program acquisition officers, selected from a defined science, technology, engineering, and mathematics (STEM)-intensive career path in the uniformed services. This dedicated line of program acquisition officers would be similar in intent, education, and experience to the Navy's engineering and aeronautical engineering duty officers. Additionally, a robust career path for USAF civilian engineers and program managers should be established that supports their critical importance to the successful execution of acquisition programs through ownership of the technical baseline. Program managers should generally be selected from the engineering and technical workforce.

The USAF supplements its organic technical workforce with on-site contractors who are not employed by the prime contractor of an acquisition program. These contractors, often hired via Advisory and Assistance Services (A&AS) contract vehicles, provide specific experience, education, certifications, and other skills to

---

[33] This recommendation is in line with the memorandum of the Under Secretary of Defense Frank Kendall in "Implementation Directive for Better Buying Power 3.0—Achieving Dominant Capabilities through Technical Excellence and Innovation." In the section "Establish stronger professional qualification requirements for all acquisition specialties," this document says "This continues the BBP 2.0 effort in this area. The DAWIA [Defense Acquisition Workforce Improvement Act] experience requirements must be supplemented to establish a stronger basis for levels of professionalism across all acquisition career fields. The Department started the Acquisition Workforce Qualification Initiative (AWQI) in BBP 2.0 to better define qualification standards. The Department is close to completing the development of experiential/proficiency standards and tasks for each of the Acquisition Career Fields by competency and competency element. This career development tool focuses on the quality versus the quantity of the experience attribute of certification and provides a higher level of measureable demonstration of experience specific to a position. AWQI demonstrated experience standards will be distributed to the Acquisition Workforce (via the Components) as a guide to assist in Talent Management with an emphasis on career development and succession planning. It will aid in developing fully qualified acquisition professionals. The Components will be responsible for their implementation methodologies" (April 9, 2015, p. 29, http://www.acq.osd.mil/fo/docs/betterBuyingPower3.0(9Apr15).pdf).

fill gaps in the Air Force's organic technical capabilities. One such vehicle for these services is the General Services Administration's (GSA's) One Acquisition Solution for Integrated Services (OASIS) contract. In a March 20, 2014, memorandum,[34] the Air Force Life Cycle Management Center (AFLCMC) commander mandated the use within AFLCMC of the Engineering, Professional, and Administrative Support Services (EPASS) Program Management Office (PMO) for all A&AS requirements. The rationale was based on economic pressures and resource constraints, and an environment that required more efficacy and innovation. The memo further directed as follows:

> The EPASS PMO will utilize the GSA OASIS [One Acquisition Solution for Integrated Services] Indefinite Quantity/Indefinite Delivery (ID/IQ) Multiple Award Contract (MAC) and will align their Labor Categories, which were developed as a best estimate of the labor required to support knowledge-based services across the federal government.

The OASIS contract uses North American Industry Classification System (NAICS)[35] codes to define labor pools. Implementation challenges seem to persist from associating the broad NAICS codes (e.g., "541330—Engineering Services") and OASIS "Pools" with the narrower EPASS labor categories (e.g., electronics engineer) and the associated specific personnel requirements. Overall, because the EPASS Program Office uses a technically acceptable, lowest evaluated price (TA/LEP) approach for selecting the contractor, OASIS task orders are awarded on a low-price basis to companies that may meet the NAICS codes but do not always have the more detailed and specific requisite skills to meet the requirements outlined by the program manager. Some PMs are provided with resources that cannot fill the defined technical gaps or meet the needs of the program.

CONCLUSION 6: OASIS task orders do not consistently meet program manager requirements and, in cases, appear to be cost-driven versus need-driven. The requirements defined by the program manager, and the technical capabilities of the personnel ultimately received, do not always align or are in conflict. This issue may reside in poorly defined requirements as provided by the program manager or in an inability to properly state requirements and fill them through the contract vehicle.

---

[34] AFLCMC memorandum, "Policy for Mandatory Use of Engineering, Professional, and Administrative Support Services (EPASS) for Advisory & Assistance Services (A&AS)," March 20, 2014.

[35] According to the U.S. Census Bureau website, "The North American Industry Classification System (NAICS) is the standard used by Federal statistical agencies in classifying business establishments for the purpose of collecting, analyzing, and publishing statistical data related to the U.S. business economy" (U.S. Census Bureau, "Introduction to NAICS," last update August 8, 2016, http://www.census.gov/eos/www/naics/).

RECOMMENDATION 7: Air Force Life Cycle Management Center (AFLCMC) leaders should work with the Engineering, Professional, and Administrative Support Services (EPASS) program management office to put in place a rigorous requirements definition process such that specific technical requirements and criteria are approved by the program manager and that contractor personnel[36] align with those requirements to meet the needs of the program. Application of AFLCMC's technically acceptable, lowest evaluated price (TA/LEP) approach should be a secondary consideration to meeting the requirement and delivering customer value.

CONTRACTING SUPPORT

The roles of the program manager and the contracting officer are clearly delineated in existing DoD policy documentation. Acquisition reform and increased oversight, however, have helped to create an unintended consequence in which the two roles are sometimes upended, leading to poorly informed or rigid implementation of some contracting methodologies. The program manager is ultimately responsible for program outcome, and as such requires appropriate support from the other members of the government acquisition team.

As discovered in both the OTB Workshop report and during the study interviews conducted by the committee, the authorities and accountabilities of the PM as they relate to the authorities and accountabilities of the contracting officer (CO) are currently causing dysfunctional, as opposed to creative, tension, which is negatively impacting government acquisition team effectiveness in the Air Force. The committee could not find documentation that specifically addresses the necessary relationship between the PM and the CO, despite the fact that DoD 5000.01[37] specifically defines the roles and authorities of the PM, and Federal Acquisition Regulation (FAR) documentation describes in clear terms the qualifications of and selection criteria for the CO. The FAR language does focus, however, on the importance of the government acquisition team, which includes both the PM and the CO, and the team's collective obligation to adhere to performance standards. DoD 5000.01 also clearly states that the PM is the designated individual in terms of responsibility and authority for meeting program cost, schedule, and performance goals and for meeting the user's operational needs. The PM, therefore, holds the ultimate responsibility for the management and technical direction of the program.

---

[36] Contractor personnel refer to services purchased to augment the organic workforce within the program office.

[37] DoD Directive Number 5000.01 provides management principles and mandatory policies and procedures for managing all acquisition programs (http://www.dtic.mil/whs/directives/corres/pdf/500001p.pdf).

The committee heard evidence that indicated the Air Force is having great difficulty rationalizing and de-conflicting certain aspects of the PM and CO roles. This has led to communication problems when selecting and executing efficient contracting strategies to meet mission requirements linked to the PM's ability to own the technical baseline. This disconnect contributes to a continuing atmosphere of distrust between the contracting and program management communities and in some cases an adversarial relationship that is not conducive to a successful technical program. Government acquisition team performance, which is critical to contract implementation and mission success, is intimately tied to the relationship between these communities and their ability to communicate.

Interviewees commented on how the Air Force contracting community's current approach to technical acquisition has fostered an atmosphere of top-down interference that increases the risk to program goals and objectives. Several PEOs and retired executives[38] commented on the historical and current state of the relationship between Air Force program managers and contracting officers, citing "egregious behavior" on the part of contracting teams, and questioning their commitment to providing the best acquisition deal possible for the Air Force program. The PEO and PM perceptions were that, rather than coming to a resolution of conflicts that allows the program to proceed, these disputes over control of the program contribute to lengthy acquisition lead times, delay in contract negotiations, and a negative impact on the implementation of critical programs.

From the highest level, and in response to an excess of reform-driven oversight, the Air Force has undergone a bifurcation of the contracting community and program management community. Two chains of authority and decision making, which may be in conflict, have arisen and have replaced an effective working relationship between contracting personnel and PMs. Where PMs once actively learned about the culture and business philosophy of the contracting community through a "boots on the ground" perspective and COs learned the culture of the program management community from being an active member of the acquisition team, in many instances there is now a significant communication barrier separating the two communities.

There are agencies, services, and Air Force product centers where this bifurcation has not been as pervasive, such as the Missile Defense Agency and the Navy, and where success can be specifically attributed to documenting and clarifying the relationship between the PM and the CO from the very beginning of the program. The committee also learned of some exemplary ways in which the PM/CO relationships were pursued in successful Navy and NASA programs, both of which were characterized by clearly defined lines of responsibility that enabled people, infrastructure, and standards to function smoothly and in concert.

[38] A full list of meeting participants can be found in Appendix C.

Program managers in the Navy and in large NASA programs typically have a technical background, but they are also versed in the art and science of program management, which are largely acquired through education, experience, and mentoring. The art aspect includes openly communicating with the COs and regarding them as valued members of the supporting team. Good relationships with contracting personnel, with responsibilities and requirements established clearly and at the beginning of the program effort, help to reduce the risk of post-contract litigation; excessive and time-consuming oversight; inaccurate or incomplete terms and conditions; and the diversion of resources for troubleshooting. Importantly, effective PEOs and PMs recognize when conflicts with the CO need to be taken to the next level, and how to escalate responses appropriately.

Lines of Authority

DoD 5000.01 establishes a clear program management line of authority (PM-LoA) that begins with the Defense Acquisition Executive (DAE) (Under Secretary of Defense for Acquisition, Technology, and Logistics) and flows down to the PM. Paragraph 3-5 of the 5000.01 regulations speaks distinctly to specific PM roles and responsibilities, and DoD Instruction 5000.02 specifically delineates the PM's role under the purview of the PEO.

FAR, DoD, and Air Force Supplements[39] delineate the contracting line of authority (CLoA), which originates with the head of agency (i.e., the Secretary of the Air Force) and flows down to the CO.

As illustrated in Figure 3.4, the lines of authority diverge at the Assistant Secretary of the Air Force (Acquisition) level. Neither the DoD 5000 series documents nor the FAR and its supplements adequately address the functional interplay between the PM and CO roles and responsibilities. In many program offices, the head contracting executive is the chief of contracts, or someone with an equivalent title. The chief of contracts is the reporting official for the CO and is responsible for ensuring that the program receives high-quality support. Although the CO's authority does not flow from the chief of contracting on a day-by-day basis, the chief sets the tone for how the designated program contracting staff supports the program. The chief also usually supports the PM's staff meetings and acts as the PM's business advisor. For purposes of this report, the term "contracting professionals" will include both the chief of contracting (if assigned) and the CO. Most PEOs and PMs understand the need for contracting professionals to provide an internal control for prudently following the appropriate contracting-related laws and policies since it is a responsibility driven by the FAR. However, the majority of acquisition decisions fall well

[39] The supplements are the *Defense Federal Acquisition Regulation Supplement* and the *Air Force Federal Acquisition Regulation Supplement*, respectively.

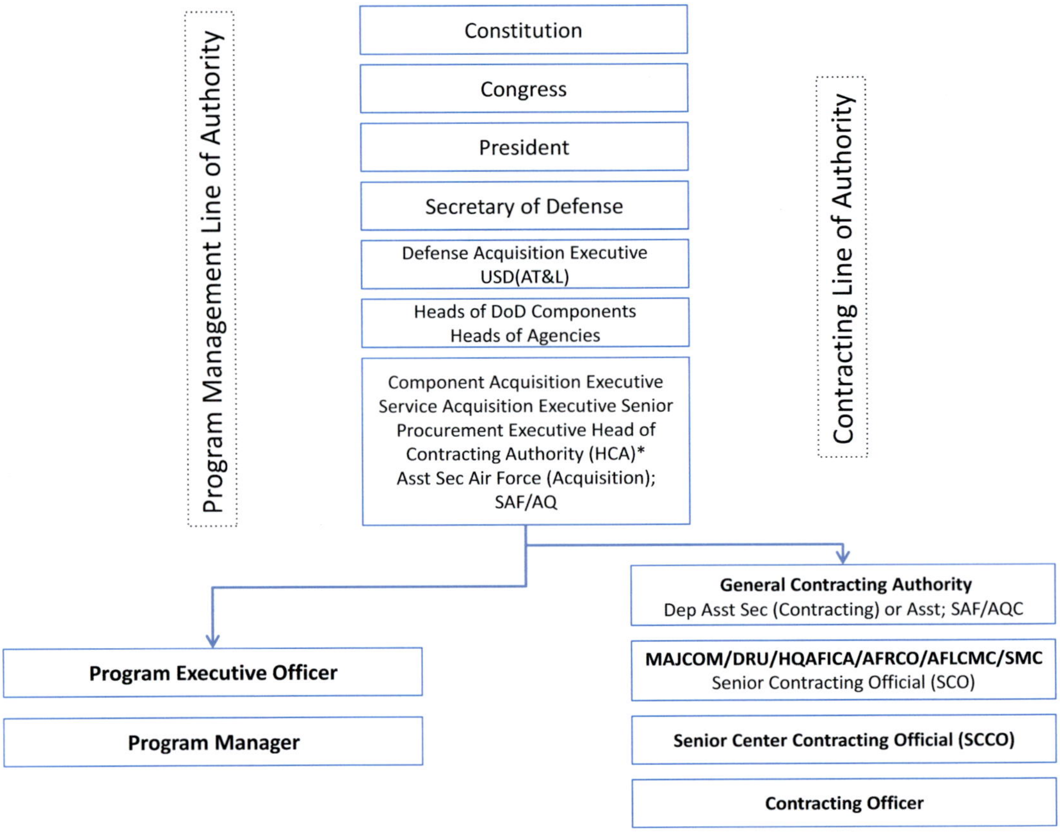

FIGURE 3.4 Illustration of program management and contracting lines of authority. NOTE: AFLCMC, Air Force Life Cycle Management Center; AFRCO, Air Force Rapid Capabilities Office; DRU, Direct Reporting Unit; HQ AFICA, Headquarters Air Force Installation Contracting Agency; MAJCOM, Major Command; SMC, Space and Missile Systems Center.

within those boundaries. Often an authority void exists regarding who has the final decision authority, which leads to dysfunctional tension, confusion, and frustration.

In the course of this study, some PEOs and PMs appeared reluctant to challenge COs on points that are clearly within the PM's purview or to escalate the issue to the CO's management. This reluctance on the part of PMs appears to stem from a lack of understanding of their authorities, or no effective escalation path for the issue, as well as past experiences gained from a culture of never challenging contracting officers. One example provided to the committee was that of a PM who was not allowed to attend Business Clearance and Contract Clearance sessions (AFFARS MP5301.9001(f)) because the CO insisted that the sessions were exclusively a contracting management chain process. The Air Force Federal Acquisition Regulation Supplement (AFFARS) requires Business Clearance Approval and Office of the Secretary of Defense Peer Review approval to (1) establish negotiation objectives for competitive procurements or (2) establish final negotiation objectives before concluding negotiations for non-competitive procurements. The committee did not find any policy prohibiting PEO or PM representation at the clearance or Peer Review sessions.

CONCLUSION 7: In many cases, there is confusion, frustration, miscommunication, and mistrust in the relationship among the COs and the PEOs and PMs they support. The committee found evidence that in some instances, contracting professionals (1) overstep their authority (for example, by dictating a specific contracting approach), (2) apply overly strict and restrictive interpretations of regulations, (3) take positions on business issues without providing adequate explanation to the rest of the government acquisition team (FAR 1.102-3), and (4) are not evaluated via their annual appraisals based on the quality of their support to the program manager, successful program execution, or meeting program office objectives within FAR guidelines.

CONCLUSION 8: The Air Force does not currently possess an adequate program management governance structure that specifies clear lines of authority, responsibility, and accountability for members of the government acquisition team. Additionally, Air Force PEOs and PMs are not mandatory participants in Business Clearance or Contract Clearance sessions, even though the negotiation objectives, key contract terms and conditions, and the nature of the deal are largely set in these sessions. Without a full awareness of contract terms and objectives, PMs and PEOs may find themselves in a position of being forced to implement an acquisition approach or execute a contractual business arrangement that they either do not understand or believe to be inappropriate.[40]

[40] Page 27 of the Under Secretary of Defense's memorandum of April 9, 2015, "Implementation Directive for Better Buying Power 3.0—Achieving Dominant Capabilities through Technical Excellence and Innovation," states under the "Improve requirements definition for services" section that "Improving services contracting requirements definition is a continuing BBP [better buying power] initiative. Defining requirements well is a challenging but essential prerequisite in achieving desired services acquisition outcomes. As most services are integrated into the performance of a mission, it is critical to get the mission owner (often an operational commander) involved in the requirement definition, as well as the acquisition and execution phases. Continuous involvement through the services acquisition phases will lead to improving requirements definition for future acquisitions" (http://www.acq.osd.mil/fo/docs/betterBuyingPower3.0(9Apr15).pdf).

RECOMMENDATION 8: The Air Force should issue a guidance memorandum that clearly specifies the lines of authority and accountability for all members of the government acquisition team.[41] This memorandum should clarify and reinforce program manager (PM) authorities and responsibilities as well as specify contracting officer responsibilities, as part of the government acquisition team, in relation to the PM. Specifically, all functional entities should provide the PM with the support necessary to attain program success. All members of the government acquisition team should be measured based on program success while complying with the law. Additionally, the Air Force should revise the Air Force Federal Acquisition Regulation Supplement to make it clear that Air Force program executive officers and PMs, or their designated representatives, are mandatory participants in business clearance and contract clearance sessions. Program management and contracting personnel should be trained in implementation of the guidance.

During the OTB workshop, several program offices expressed frustration that the contracting office is not accountable for program success or failure and is focused on aspects that are taken out of context with the overall program. This lack of program accountability creates a disparity in incentives between the PM, who needs the technical support, and the CO, who is focused on process issues and timelines. COs who are not held accountable for program success, yet have the authority to constrain the PM from hiring the best engineering talent to support the program, often use cost control as the paramount metric, thus subverting the necessary balance among cost, schedule, and technical scope. CO mandates that a specific contract vehicle type be used (e.g., lowest price, technically acceptable [LPTA] for acquiring S&E support) can preclude the PM from hiring highly skilled engineering talent (and have done so). This relationship is a pressing issue for the Air Force to address in order to give the PM access to the technical expertise necessary to own the technical baseline.

---

[41] This recommendation is in line with the Under Secretary of Defense's memorandum, "Implementation Directive for Better Buying Power 3.0—Achieving Dominant Capabilities through Technical Excellence and Innovation," under the section "Eliminate Unproductive Processes and Bureaucracy" and the subsection "Emphasize acquisition chain of command responsibility, authority, and accountability." This section states "This initiative is a continuing effort from BBP 2.0. The chain of command for acquisition programs runs upward from the [program manager] through the [program executive officer] to the [component acquisition executive] and, for [Acquisition Category] ACAT I, ACAT IA, and other programs so designated, to the [defense acquisition executive]. The responsibility and authority for program management, to include program planning and execution, is vested only in these individuals. Staff and other organizations provide support to this chain of command" (April 9, 2015, p. 18, http://www.acq.osd.mil/fo/docs/betterBuyingPower3.0(9Apr15).pdf).

CONCLUSION 9: Not all government acquisition team members are currently accountable for program progress, success, or failure, and the primary objectives and requirements of the program manager and contracting officer are disconnected or in some cases appear to be in opposition to one another, creating disparity in effort and incentives.

RECOMMENDATION 9: Contracting professionals' appraisals should have appropriate objectives and metrics tied directly to the program office or organization's mission success. The program executive officer or the program manager or their designee should be required to provide written performance input to the contracting professionals' annual appraisals. Contracting professionals should engage with the program office and be well trained and experienced with their accountability and responsibility for delivering support to the assigned Air Force organization and mission.

The above issues between program officers and contracting officers have contributed to an ongoing erosion of trust and, in several cases, an adversarial relationship between the PM and CO, which has proven highly detrimental both to the acquisition process and to meeting mission needs.

Lowest Price, Technically Acceptable

During the data-gathering meetings, the committee heard several examples of dissatisfaction with the indiscriminate use of the LPTA source selection as a contracting methodology. LPTA is a process that is one of the methodologies contained in what FAR 15.101 describes as the "Best Value Continuum," defined by the FAR as follows:

> An agency can obtain best value in negotiated acquisitions by using any one or a combination of source selection approaches. In different types of acquisitions, the relative importance of cost or price may vary. For example, in acquisitions where the requirement is clearly definable and the risk of unsuccessful contract performance is minimal, cost or price may play a dominant role in source selection. The less definitive the requirement, the more development work required, or the greater the performance risk, the more technical or past performance considerations may play a dominant role in source selection.

"Better Buying Power 2.0" contained the following guidance:

> When LPTA is used, define Technically Acceptable to ensure needed quality: Industry has expressed concerns about the use of Lowest Price, Technically Acceptable (LPTA) selection criteria that essentially default to the lowest price bidder, independent of quality.

> Where LPTA is used, the Department needs to define TA appropriately to ensure adequate quality.[42]

During the OTB workshop, participants noted instances of inappropriate use of LPTA, where the government acquisition team would have been better served by using a more integrated and best-value approach to enhance the ability to tradeoff between non-cost factors to ensure the selected contractor could meet requirements. LPTA is an evaluation for a specific point in time and does not possess metrics for forecasting impacts at various program stages. Another major concern heard during interviews for both the OTB workshop and this study was that the LPTA methodology does not allow for consideration of the specific engineering expertise, skills, and experience criteria needed to provide high-quality and specific technical resources—items paramount to owning the technical baseline. It appeared that some of the existing contracts and task orders awarded using LPTA methodology were written to accommodate a larger group of contract awardees, including small businesses, and to save money through a simpler source selection process and artificially suppressed contractor labor rates, as opposed to keeping the focus on quickly and efficiently hiring proficient, competent technical contractor resources.

A highly technical organization such as NASA does in fact employ contracting vehicles that are as simple as a one-page letter of agreement, characterized by a director-level contracting team engaged in long-term strategizing, case-by-case intellectual property sharing, and joint cost-sharing activities. Programs need flexibility to adopt a more nuanced approach than LPTA to provide unique solutions for solving technical problems.

> **CONCLUSION 10:** The LPTA contract type was not intended to be mandatory or for the acquisition of all technical products and systems, but it has evolved in that direction in the current acquisition climate. Indiscriminate use of LPTA as a proposal evaluation and contractor selection methodology has resulted in poor outcomes and frustrated program managers, who do not receive the required high-quality technical support that is required for understanding and owning the technical baseline. When not used properly, LPTA can result in a lack of technical and engineering expertise. It can lead to an inadequate mix of talents for the contract, cause long delays in the process of contract execution, and create excessive turnover in the workforce owing to low wages.

---

[42] Under Secretary of Defense, "Better Buying Power 2.0: Continuing the Pursuit for Greater Efficiency and Productivity in Defense Spending," Memorandum for Defense Acquisition Workforce, November 13, 2012, http://bbp.dau.mil/doc/USD-ATL%20Memo%2013Nov12%20-%20BBP%20 2.0%20Introduction.pdf, p. 5.

RECOMMENDATION 10: The Assistant Secretary of the Air Force for Acquisition should clarify the criteria for use of the lowest price, technically acceptable (LPTA) methodology and ensure there are avenues for the government acquisition team to discuss its appropriateness for meeting mission requirements. LPTA should not be applied to complex, multiyear, multidiscipline programs or knowledge-based service contracts that require high-end acquisition and technical talent. A decision to use LPTA should depend on clear and unambiguous requirements, underlying market research, and relevant information acquired during government and contractor interactions, such as "industry days." If there is a requirement that demands special treatment, the case should be made in the requirements definition, acquisition strategy, and pre-request for proposal activity.

In the case of contracting, OTB will enable the government acquisition team to make smart decisions, facilitate contracting trade-offs, and enhance its overall ability to implement the contract. Regaining the OTB could enable the USAF to carry out contracting activities in an efficient and timely manner through informed decision making.

FUNDING

In a constrained budgetary atmosphere, the efficient use of available funds to support weapons systems is paramount to meeting mission requirements. Air Force leaders have recognized this need and have already begun to employ more flexible means of funding acquisition staff. Of even greater concern, however, is the need to recognize that the Air Force cannot accept new programs without the ability to execute them.

As was initially discussed during the OTB workshop and reaffirmed throughout the study, the type of funds used for personnel is a key issue. Funding is a key variable for hiring top-shelf engineers from academia or industry and supporting their advanced education and training in core competencies and providing mentors to guide their careers through increasing levels of responsibility such that these engineers fulfill the needs of PEOs and PMs charged with executing programs. A lack of adequate and timely funding will continue to limit the ability of acquisition center functionaries to create a workforce that is capable of meeting the high technical demands of Air Force weapon systems.

The committee heard evidence that the Air Force was already in the process of realigning funding for development of the acquisition workforce.[43] The use of

[43] Appendix D contains a memorandum from the Commander of AFLCMC to the study committee that highlights current activities at AFLCMC.

research, development, testing, and evaluation (RDT&E) funding ("3600" funds) instead of operations and maintenance (O&M) funding ("3400" funds) is already in the process of being used to fund the acquisition workforce. This move, it is widely believed, will allow more flexibility in the hiring and training of the organic engineering workforce. The use of 3600 funding would allow civil service engineers to be secured, trained, and employed in support of program office needs and to fill the necessary gaps.

CONCLUSION 11: Lack of adequate and timely funding limits the ability of acquisition-center functional leads to shape the workforce to meet the demand for knowledgeable and experienced technical talent.[44]

RECOMMENDATION 11: The Air Force should complete the shift from operations and maintenance funds to research, development, testing, and evaluation funds for funding acquisition staff. Additionally, the Assistant Secretary of the Air Force for Acquisition should require program managers to include in their program financial plan such a budget, as necessary, to fully fund the in-house technical effort.

The committee heard additional evidence of other Air Force efforts to address funding issues that relate to the Air Force's ability to own the technical baseline of its acquisition programs. One such effort was the pursuit by the Air Force to utilize the flexibility provided by the Air Force Working Capital Fund (AFWCF).[45] The AFWCF would facilitate PM use of in-house engineering staff to facilitate the technical work necessary to own the technical baseline. This would be done by expanding the use of the AFWCF from its current applications in logistic support. While the committee did not receive enough information on the Air Force's current efforts to utilize the AFWCF to make a recommendation, the committee notes its value to owning the technical baseline. The committee also heard evidence of problems arising from a lack of proper resourcing, as required by the Weapons Systems Acquisition Reform Act (WSARA) of 2009, of new major defense acquisition programs (MDAPs).[46] Starting new programs without adequate funding significantly hampers the Air Force's ability to own the technical baseline of its programs. Budget cuts to established MDAPs during the yearly planning, pro-

[44] This conclusion was also a major recurring theme in National Research Council, *Owning the Technical Baseline for Acquisition Programs in the U.S. Air Force: A Workshop Report*, The National Academies Press, Washington, D.C., 2015, p. 4.

[45] General Ellen M. Pawlikowski, Commander, Air Force Materiel Command, interview with the committee on March 23, 2016.

[46] Lynn M. Eviston, Director of Plans and Programs, Air Force Life Cycle Management Center, interview with the committee on March 31, 2016.

gramming, budget, and execution (PPBE) process can cause the Air Force to lose ownership of the technical baseline in programs where it once had had ownership. The Air Force, as highlighted previously in the Workforce Management section, is over-programmed. Inadequate funding for programs results in elongated delivery schedules and difficulties in achieving cost, schedule, and performance goals.

CONCLUDING REMARKS

Tomorrow's wars are being fought in today's program development offices. To win those wars, urgent and dramatic steps are needed to ensure that the Air Force removes several major barriers to success. Owning the technical baseline is a critical component of the Air Force's ability to regain and maintain acquisition excellence. There are definitely very professional PEOs and PMs who are performing well in spite of the barriers, but it is clear that the USAF needs to take immediate steps to emphasize the value of its acquisition professionals, ensure sustained leadership within the acquisition community, reinforce the PM's authority and accountability, clarify the role of the contracting officer vis-à-vis the PM, strengthen and expand the technical knowledge base and expertise of the acquisition workforce, and continue to eliminate barriers and avoid creating new ones. These necessary steps for owning the technical baseline are especially important in light of the shorter and shorter time frames within which the Air Force needs to develop and deploy warfighting capabilities to meet rapidly emerging and changing threats.

Appendixes

A

Statement of Task

An ad hoc committee of interdisciplinary experts will be formed to:

1. Identify the strategic value to the U.S. Air Force in properly controlling, as well as the risk of not controlling, the technical baselines of its programs.
2. Investigate how others (e.g., services, government agencies, and commercial industry) control technical baselines and provide an assessment on the most promising mechanisms for potential application within the U.S. Air Force.
3. Recommend ways to remove or remediate barriers across the U.S. Air Force, such as barriers identified in Owning the Technical Baseline for Acquisition Programs in the U.S. Air Force: A Workshop, to properly control the technical baselines of future programs.
4. Recommend ways for the U.S. Air Force to assess and adopt any identified methods for controlling the technical baselines across its acquisition programs.

B

Biographical Sketches of Committee Members

HENRY A. OBERING III, *Chair*, is a Booz Allen Hamilton executive vice president. An expert in acquisition and program management, he works with aerospace clients in the Air Force and the National Aeronautics and Space Administration (NASA). Prior to joining Booz Allen Hamilton, he led a comprehensive review of the National Reconnaissance Office for the Director of National Intelligence that provided a new charter for that organization. Mr. Obering retired from the U.S. Air Force (USAF) as a lieutenant general with more than 35 years of experience in space and defense systems development, integration, and operations. He served as director of the 8,500-person Missile Defense Agency (MDA). He was the Department of Defense (DoD) acquisition executive for the nation's $10 billion per year missile defense portfolio. In addition, he served as the program manager for the Ballistic Missile Defense System. Prior to his assignment at MDA, he planned and programmed 68 joint Air Force and international programs with a $28 billion budget as mission area director for information dominance on the Air Staff. Mr. Obering entered the Air Force in 1973 after completing the University of Notre Dame's ROTC program as a distinguished graduate. He received his pilot wings in 1975 and flew operational assignments in the F-4E. Later, he was assigned to the Space Shuttle Program and participated in 15 space shuttle launches as a NASA orbiter project engineer and was responsible for integrating firing room launch operations. Other assignments include tours with the Air Force Inspector General, the Defense Mapping Agency, and the Electronic Systems Center. Mr. Obering has twice earned DoD's highest noncombat award, the Defense Distinguished Service Medal for leadership. He was honored by the National Defense Industrial Association's

Missile Defense Division with the 2011 Kadish Award for Acquisition Excellence. He received an M.S. in astronautical engineering from Stanford University and a B.S. in aerospace engineering from the University of Notre Dame.

LAWRENCE J. DELANEY, *Vice Chair*, retired as chairman of the board of trustees of Clarkson University in October 2015. He has been the acting Secretary and acting Under Secretary of the Air Force. Dr. Delaney also served as the Assistant Secretary of the Air Force for Acquisition, as well as the Air Force's service acquisition executive, responsible for all Air Force research, development, and acquisition programs and activities and chief information officer of the Air Force. He provided direction, guidance, and supervision of all matters pertaining to the formulation, review, approval, and execution of acquisition plans, policies, and programs. He has more than 40 years of international experience in high technology program acquisition, management, and engineering, focusing on space and missile systems, information systems, propulsion systems, and environmental technology. He retired as the executive vice president of operations and president of the advanced systems development sector of Titan Corporation. Previously, he held distinguished positions with Arete Associates, Inc.; Delaney Group, Inc.; BDM Europe, Berlin; and the environmental and management systems group at IABG, Ottobrunn, Germany. Dr. Delaney earned his Ph.D. in chemical engineering at the University of Pennsylvania. For the National Academies of Sciences, Engineering, and Medicine, he served as vice chair of the Board on Army Science and Technology, chair of the Air Force Studies Board, and vice chair of the Army Science Board.

DONALD R. ERBSCHLOE is a technical consultant working with diverse customers such as NATO, NASA, the Air Force Research Laboratory, and government/industry partnerships in Silicon Valley. He recently retired as the chief scientist of the Air Mobility Command (AMC), Scott Air Force Base, Illinois. AMC provides global reach for the U.S. Air Force, with a fleet of nearly 1,200 aircraft and 130,000 personnel worldwide. He was the 2014 recipient of the Harold Brown Award, the Air Force's top science and technology recognition. Prior to this assignment, Dr. Erbschloe was the chief operating officer of the Department of Energy's Office of Science, which manages 10 world-class laboratories and is the single largest supporter of basic research in the physical sciences in the United States. Previously, Dr. Erbschloe had a 28-year military career in the Air Force. His career was balanced among flight operations, academia, and scientific and technical management. He was a command pilot with 4,000 flying hours in a variety of aircraft. He served three tours on the faculty at the Air Force Academy in the Department of Physics and as the director of faculty research. He was the military assistant to three Air Force chief scientists at the Pentagon. His final active duty assignment was as the commander and deputy director of the Air Force Office of Scientific Research in

Arlington, Virginia. He earned his doctorate in physical electronics from Oxford University.

MILLARD S. FIREBAUGH is a Glenn L. Martin Institute Professor of Practice at the University of Maryland, having retired with the rank of rear admiral after more than 33 years of service with the U.S. Navy as an engineering duty officer. He became vice president of innovation at General Dynamics, Electric Boat Division, from which he retired in 2005. His area of technical interest is naval engineering, principally the design, systems engineering, naval architecture, and integration of weapons and combat systems of warships, with particular emphasis on nuclear-powered submarines. His focus is on the translation of naval requirements into detailed engineering, design, and construction methodology for complex warships, with specific interest in the hydrodynamics and hydroacoustics of submarines relating to powering and propulsion, quieting, sensor system performance, weapon launch, and integration of features. Admiral Firebaugh is a member of the National Academy of Engineering, and he served on the National Academies' Committee on Materials Research for Defense-After-Next, its Panel on Structural and Multi-functional Materials, and the Naval Studies Board study that authored *Mainstreaming Unmanned Undersea Vehicles into Future U.S. Naval Operations*. He received a B.S. in physics, an M.S. in electrical engineering, a naval engineering degree, and a D.Sc. in ocean engineering from the Massachusetts Institute of Technology (MIT).

MICHAEL D. GRIFFIN is the chairman and chief executive officer of Schafer Corporation, a leading provider of scientific, engineering, and technical services and products in the national security sector. He was previously a King-McDonald Eminent Scholar and professor of mechanical and aerospace engineering at the University of Alabama, Huntsville, was the Administrator of NASA from 2005 to 2009, and prior to that was the Space Department head at the Johns Hopkins University (JHU) Applied Physics Laboratory (APL). He has also held numerous executive positions in industry, including president and chief operating officer of In-Q-Tel, CEO of Magellan Systems, general manager of Orbital Science Corporation's Space Systems Group, and executive vice president and chief technical officer at Orbital. Dr. Griffin's earlier career includes service as both chief engineer and associate administrator for exploration at NASA and as the deputy for technology at the Strategic Defense Initiative Organization (SDIO). Prior to joining SDIO in an executive capacity, he played a key role in conceiving and directing several first-of-a-kind space tests in support of strategic defense research, development, and flight testing. These included the first space-to-space intercept of a ballistic missile in powered flight, the first broad-spectrum spaceborne reconnaissance of targets and decoys in midcourse flight, and the first space-to-ground reconnaissance of ballistic missiles during the boost phase. He also played a leading role in other space missions in earlier work at the JHU Applied Physics Laboratory, NASA's Jet Propulsion Laboratory, and the Computer Science Corporation. Dr. Griffin was an adjunct professor for 13 years at the University of Maryland, JHU, and George Washington University, offering courses in spacecraft design, applied mathematics, guidance and navigation, compressible flow, computational fluid dynamics, spacecraft attitude control, astrodynamics, and introductory aerospace engineering. He is a registered professional engineer in Maryland and California, and is the lead author of more than two dozen technical papers and the textbook *Space Vehicle Design*. Dr. Griffin is a member of the National Academy of Engineering and the International Academy of Astronautics, an honorary fellow and the current president of the American Institute of Aeronautics and Astronautics (AIAA), a fellow of the American Astronautical Society, and a senior member of the Institute of Electrical and Electronic Engineers. He is the recipient of numerous honors and awards, including the NASA Exceptional Achievement Medal, the AIAA Space Systems Medal and Goddard Astronautics Award, the National Space Club's Goddard Trophy, the Rotary National Award for Space Achievement, the Missile Defense Agency's Ronald Reagan Award, and the DoD Distinguished Public Service Medal, the highest award that can be conferred on a nongovernment employee. He received his Ph.D. in aerospace engineering from the University of Maryland and has been recognized with honorary doctoral degrees from Florida Southern College and the University of Notre Dame.

GARY A. KYLE is president and CEO of Persistent Agility, Inc. (PAI), a veteran-owned small business providing unique, innovative, and proven business and mission solutions. He leads high-level teams that independently advise C-level executives and U.S. government senior leaders on complex acquisition, contracting, business, and strategy matters. Mr. Kyle served as chairman of an ad hoc committee that assessed and made strategic business, political, and technical recommendations for a $10 billion national security space system. He was the lead consultant on a $13.7 billion space system proposal development for a critical national asset, ensuring the client understood key customer requirements. Prior to his present position, Mr. Kyle served as a principal at Booz Allen Hamilton, where he led a team of 52 geographically dispersed business analytics consultants. His team provided acquisition, contracting, cost, and financial expertise for Air Force clients at the Pentagon (HQ Air Force Financial Management), AMC, and Air Force Space Command. Mr. Kyle was hand selected by the Air Force Space Command Commander and the National Reconnaissance Office Director as the Evolved Expendable Launch Vehicle Tiger Team business lead. In this role, he collaborated with a cross-functional team of technical, operational, financial, and contracting executives to define the future acquisition tenets for the $72 billion space launch vehicle program. Mr. Kyle orchestrated a 60-person team that analyzed the Air Force chief information officer's enterprise-wide information technology efficiency initiatives and recommended savings valued at $800 million. Mr. Kyle holds an M.A. in telecommunications from the George Washington University; an M.S. in national resource strategy from National Defense University; and an M.S. in procurement and acquisition management from Northrop University. He is a graduate of both the Defense Systems Management College Program Management Course and Senior Acquisition Course and holds the highest DoD acquisition certifications in both program management and contracting.

THOMAS L. MAXWELL is an independent aerospace consultant and has a B.S. from Purdue University, an M.S. from the University of Cincinnati, and an M.B.A from Xavier University. He joined GE in 1969 and has held positions in engine test, commercial turbine design, product support, sales, product management, and engineering leadership. He has spent the last 30 years working for U.S. and international military programs. His most recent assignments were as director of the F110 Program, director of the F136 Program, and general manager of Military Propulsion Engineering. In this assignment he had engineering responsibility for 25,000 installed engines and GE propulsion development programs for ship, industrial gas turbine derivatives, and both fixed and rotary wing aircraft. Mr. Maxwell was recognized in 2012 as a Distinguished Engineering Graduate of Purdue University. He has served on the Engine Independent Review Team chartered by the USAF propulsion Program Executive Office, on Purdue engineering advisory committees, and on the boards of trustees of charitable organizations.

SUE C. PAYTON is president of SCI Aerospace, Inc. She is a former Assistant Secretary of the Air Force for Acquisition and former Deputy Under Secretary of Defense. Ms. Payton served as a presidential appointee in two presidential administrations with responsibility for acquisition, procurement, and technology transition. In the Air Force, she directed an annual budget in excess of $30 billion that included major weapon systems such as unmanned aircraft, fighter aircraft and munitions, information technology, alternative energy, advanced manufacturing, and intelligence programs. Prior to her government service in 2001, she worked in senior leadership capacities at Lockheed Martin/Martin Marietta with responsibility for leading DoD initiatives in areas such as information technology, complex space systems, intelligence, and operations. Her program achievements have been noted on numerous television shows and dozens of media publications, and she is the inaugural recipient of the DoD 2011 Manufacturing Technology Champion Award. In her role as the president of SCI Aerospace, she serves on industry and research laboratory advisory boards. A graduate of Eastern Illinois University (EIU), she earned an M.S. in systems acquisition management and computer information systems from the University of Southern California and Nova Southeastern

University and received an honorary Ph.D. in public service from Eastern Illinois University. She is a member of many government/industry/educational/nonprofit organizations, including the EOD Warrior Foundation, U.S. Geospatial Intelligence Foundation, the EIU Foundation board of directors, the Air Force Academy Systems Engineering Program advisory council, Purdue University's President's Council, Southern Methodist University's Caruth Institute for Entrepreneurship, and the Doolittle Institute.

RICHARD T. ROCA is director emeritus at JHU APL. He became director at JHU APL in January 2000 and stepped down from that position in June 2010. While director, he led a not-for-profit University Affiliated Research Center that performs research and development work on behalf of the DoD, primarily the U.S. Navy, and for NASA and other federal agencies. Currently, Dr. Roca is a member of the board of directors of the Charles Stark Draper Laboratory, Inc., and a member of the board of trustees of Olin College of Engineering. Dr. Roca spent the first years of his professional career with the AT&T Corporation. He joined Bell Laboratories to design communications equipment and networks and rose over three decades to major corporate leadership roles there. He is a fellow of the American Society of Mechanical Engineers and a past vice president of its Board on Engineering Education. He is a visitor for the Accreditation Board for Engineering and Technology, which accredits U.S. engineering schools. Dr. Roca received a Sc.D. from MIT in mechanical engineering. In 1977, he was awarded a congressional fellowship and worked on the National Energy Act in the House Commerce Committee's Subcommittee on Energy and Power. In 2011, he received the Secretary of Defense Award for Outstanding Public Service.

WILLIAM J. STRICKLAND is president and CEO of the Human Resources Research Organization (HumRRO) in Alexandria, Virginia. Before his appointment as CEO, he spent more than 10 years as a HumRRO vice president, directing its Workforce Analysis and Training Systems Division. Before joining HumRRO, he served in the USAF and retired with the rank of colonel; in his last assignment, he was the director for Air Force human resources research. He is a fellow of the American Psychological Association (APA), past president of its Division of Military Psychology, and served for 6 years as that division's representative on the APA Council of Representatives. Dr. Strickland currently serves as a member-at-large on the APA board of directors, and has represented APA on the Board of the Consortium of Social Science Organizations and on the Council of the Federation for Brain and Behavioral Sciences. He has been a member of five previous National Academies committees. He is a graduate of the U.S. Air Force Academy and earned a Ph.D. in industrial and organizational psychology from Ohio State University.

DEBORAH L. WESTPHAL is managing director of the strategy advisory firm Toffler Associates. Recognized globally for her expertise in strategy, innovation, and organizational transformation, Ms. Westphal helps organizations understand the forces that drive change in their industries and the world and identifies the best courses of action to create enduring success. Ms. Westphal came to Toffler Associates in 1999 after 13 years as a senior government official in the USAF. Her work in the area of technology and advanced concepts for air vehicles, missiles, and space systems has been recognized with numerous awards, including from the California Air Force Association (AFA), a USAF Meritorious Civilian Award, an AFA Los Angeles chapter Civilian of the Year award, and an AFA Medal of Merit. Ms. Westphal has also served on the U.S. Army Science Board, the board of directors of the Greater Los Angeles Chapter of the National Defense Industrial Association and the board of directors of the Schriever Chapter 147 of the AFA. Currently, Ms. Westphal serves on the Air Force Studies Board of the National Academies. As managing director of Toffler Associates since 2007, she is an acknowledged expert in the aerospace industry and brings a wealth of experience in a wide range of other sectors, including materials, transportation, security, space, hospitality, and telecommunications, as well as U.S. defense, intelligence, and civilian government. Ms. Westphal's success can be traced to her unique combination of education and experience. She holds a B.S. in electrical engineering and an M.B.A. from Webster University and has completed executive education at Harvard Business School and the Wharton School of Business.

REBECCA WINSTON is president of Winston Strategic Management Consultants. She is a former chair of the board of the Project Management Institute (PMI). An experienced and recognized expert, she was a fellow in PMI and was named one of the 100 most influential experts globally on the subject of project and program management (PM) in the fields of research and development (R&D), energy, environmental restoration, and national security, and is well known throughout the United States and globally as a leader in the PM professional world. Ms.Winston has more than 25 years of experience in program and project management, primarily on programs funded by the U.S. government. She is a graduate of the University of Nebraska's College of Law, Juris Doctorate (1980), in Lincoln, Nebraska. She has a B.S. in education from Nebraska Wesleyan University and an M.S. in biology from Idaho State University. She is a licensed attorney in Iowa and Nebraska. Active in PMI since 1993, Ms. Winston helped pioneer PMI's specific interest groups (SIGs) in the 1990s, including the Project Earth and Government SIGs, and was a founder and first co-chair of the Women in Project Management SIG. She served two terms on the PMI board of directors as director-at-large, secretary treasurer, vice chair (for 2 years), and chair (2002). She was elected a PMI fellow in 2005. She is also a member of the American Bar Association and the Association of Female Executives in the United States. Ms. Winston currently serves as a consultant to organizations such as the U.S. National Nuclear Security Administration, the Department of Energy, and the Department of Homeland Security, as well as many private enterprises, on topics ranging from program and project management to project reviews, risk management, and vulnerability assessments. She has extensive recent PM experience in the areas of national defense and security and has worked closely with local, regional, and national officials, including Congress and the Pentagon.

C

Meetings and Speakers

MEETING 1
January 13-15, 2016
Keck Center of the National Academies of Sciences, Engineering, and Medicine
Washington, D.C.

Gordon England, former Deputy Secretary of Defense and former Secretary of the Navy
William H. Gerstenmaier, SES, Associate Administrator for Human Exploration and Operations, NASA
Lee Buchanan, President and CEO, Arete Associates
Vice Admiral James D. Syring, Director, Missile Defense Agency
Robert B. Raines, SES, Associate Administrator for Acquisition and Project Management, NNSA
Jaiwon Shin, SES, Associate Administrator for Aeronautics, NASA
Stan Soloway, President and CEO, Professional Services Council
Edward J. Hoffman, Chief Knowledge Officer, NASA
Lt. Gen. Ron Kadish (USAF, retired), former Director, Missile Defense Agency

MEETING 2
February 8-10, 2016
Keck Center of the National Academies of Sciences, Engineering, and Medicine
Washington, D.C.

Vice Admiral David A. Dunaway (USN, retired), former Commander of Naval Air Systems Command (NAVAIR)
Jacques Gansler, professor emeritus, School of Public Policy, University of Maryland; former Under Secretary of Defense for AT&L
Lt. Gen. Charles R. "CR" Davis (USAF, retired), President & CEO, Seabury Global Aerospace and Defense
Richard B. Clifford Jr., SES, Deputy General Counsel for Acquisition, Headquarters U.S. Air Force
James P. Woolsey, SES, President, Defense Acquisition University
Blaise J. Durante, Director, Blaise J. Durante & Associates, Inc.
Ronald Poussard, Executive Vice President for business development and contracts, Advanced Concepts and Technologies International (ACT-I)
Col. George N. Schwartz, 645 AESG/CC
John Weiler, Vice Chair, Interoperability Clearinghouse and IT Acquisition Advisory Council

MEETING 3
March 30-April 1, 2016
Courtyard Dayton Beavercreek
Dayton, Ohio

General Ellen Pawlikowski, Commander, Air Force Materiel Command (AFMC)
Douglas L. Loverro, SES, Deputy Assistant Secretary of Defense for Space Policy
Thomas D. Robinson, SES, Director of Contracting, AFLCMC
Jorge F. Gonzalez, SES, Director, Engineering and Technical Management/ Services Directorate, AFLCMC
Col. Keith L. Bearden, Deputy Director, Engineering and Technical Management, AFMC
Col. Steven J. Butow, Defense Innovation Unit Experimental (DIUx)
Kevin W. Buckley, SES, PEO of Air Force Mobility Programs, AFLCMC
Brig. Gen. Eric T. Fick, PEO for Fighters and Bombers, AFLCMC
Gary L. Poleskey, Vice President, Dayton Aerospace, Inc.
Kathy L. Watern, SES, Deputy Assistant Secretary for Cost and Economics, SAF/FMC
Lt. Gen. Samuel A. Greaves, Commander, Space and Missile Systems Center, Air Force Space Command
Gwynne Shotwell, President and COO, SpaceX

Lynn M. Eviston, Director, Plans and Programs, AFLCMC
Brig. Gen. Duke Z. Richardson, PEO for Tankers, AFLCMC
Lynda T. Rutledge, SES, PEO and Director for the Agile Combat Support Directorate, AFLCMC
Christine H. Fox, Senior Advisor, Johns Hopkins Applied Physics Laboratory; former Acting Deputy Secretary of Defense and former Director of Cost Assessment and Program Evaluation (CAPE)

MEETING 4
May 16-18, 2016
The National Academy of Sciences
Washington, D.C.

This final meeting was a committee-only working meeting.

D

Memorandum from the Air Force Life Cycle Management Center

DEPARTMENT OF THE AIR FORCE
HEADQUARTERS AIR FORCE LIFE CYCLE MANAGEMENT CENTER
WRIGHT-PATTERSON AIR FORCE BASE OHIO

MEMORANDUM FOR AIR FORCE STUDIES BOARD

20 JUN 2016

FROM: AFLCMC/CC
1865 4th Street
WPAFB, OH 45433

SUBJECT: OWNING THE TECHNICAL BASELINE

1. The AF Life Cycle Management Center since March 2016 is extensively data mining gaps, seams and shortfalls in its life cycle acquisition workforce from both capacity and capability aspects. The capacity issue must first be fixed in order for us to realistically focus on the capability gaps within the workforce. Although this study is primarily focused on the engineering career field, in order to effectively execute the concept of "owning the technical baseline", we must also recognize the need for contracting, financial management and logisticians, key enablers which are also significantly under-resourced.

2. From an engineering aspect, we are using several tools under our Strategic Resource Management (SRM) concept to analyze and fully understand where we can accept risk, the level of risk we are today and options to mitigate. The Human Capital Competency Management Tool, another initiative under SRM, allows us to fully understand the level of our workforce and the associated gaps. The data on the attached pages highlights this information for only the engineering workforce within AFLCMC.

3. From a strategic perspective, using SRM data, we were effectively able to articulate to the AF Corporate Structure the need to fund the current level of civilian workforce to ~98.5% across the FYDP allowing us to fill a significant number of forced vacancies. Also, in this same time frame (FY18-FY22) tied to the FY18 POM, the AF will request OSD to allow us to move our workforce payroll from a 3400/O&M to a 3600/RDT&E based appropriation enabling greater transparency to this area. The AF also provided AFLCMC with 144 additional positions to execute the Advanced Pilot Training program, UH-1N and KC-46 PSI efforts, but this still only funds these programs to about 40% of the total requirement.

4. Tied into the FY19 Planning choices, we are also requesting a move of this workforce to a reimbursable structure in order to tie the workload to the resources needed to execute the programs. Although not a short term solution, we believe over the long-term we should be able to size the workforce to the necessary levels needed to effectively execute the mission. However, without this approval and additional resources needed, our organization stays at high risk.

5. If there is any questions or concern regarding the information within, my POC is Ms Lynn Eviston, AFLCMC/XP, lynn.eviston@us.af.mil, (937) 904-7242.

JOHN F. THOMPSON
Lieutenant General, USAF
Commander

Attachment:
Supporting Documentation